AN 01-85FGD-1

FLIGHT HANDBOOK

Navy Models

F9F-6, -6P AIRCRAFT

Schiffer Military/Aviation History
Atglen, PA

The U.S. Navy F9F-6, -6P flight handbook used to print this book is from the collection of Michael L. Roberts. We wish to thank Commander L.N. Carota, USN, Technical Documentation Policy and Programs Office, for his cooperation in ensuring that this manual was cleared for release.

Cover photography from the Michael L. Roberts collection.

Library of Congress Catalog Number: 94-73978

Printed in China.
ISBN: 0-88740-778-1

Published by Schiffer Publishing Ltd.
77 Lower Valley Road
Atglen, PA 19310
Please write for a free catalog.
This book may be purchased from the publisher.
Please include $2.95 postage.
Try your bookstore first.

AN 01-85FGD-1

Flight Handbook

NAVY MODELS

F9F-6, -6P
AIRCRAFT

THIS PUBLICATION SUPERSEDES AN 01-85FGD-1
DATED 15 JULY 1954 REVISED 15 NOVEMBER 1954

PUBLISHED UNDER AUTHORITY OF THE SECRETARY OF THE AIR FORCE
AND THE CHIEF OF THE BUREAU OF AERONAUTICS

15 April 1956

LIST OF REVISED PAGES ISSUED

INSERT LATEST REVISED PAGES. DESTROY SUPERSEDED PAGES

NOTE: The portion of the text affected by the current revision is indicated by a vertical line in the outer margins of the page.

*The asterisk indicates pages revised, added or deleted by the current revision.

BuAer

ADDITIONAL COPIES OF THIS PUBLICATION MAY BE OBTAINED AS FOLLOWS:

USAF ACTIVITIES.—In accordance with Technical Order No. 00-5-2.

NAVY ACTIVITIES.—Submit request to nearest supply point listed below, using form NavAer-140; NASD, Philadelphia, Pa.; NAS, Alameda, Calif.; NAS, Jacksonville, Fla.; NAS, Norfolk, Va.; NAS, San Diego, Calif.; NAS, Seattle, Wash.; ASD, NSC, Guam.

For listing of available material and details of distribution see Naval Aeronautics Publications Index NavAer 00-500.

TABLE OF CONTENTS

NAVY

IMPORTANT

To gain the maximum benefits from this handbook it is imperative that you read this page carefully.

FOREWORD

The function of this flight handbook is to acquaint the pilot with the airplane, furnishing information necessary for normal and emergency flight. All operating procedures in this book are based on actual flight tests.

It is most important that the pilot keep abreast of all pertinent Interim Revisions which may have been issued since the publication of this handbook, but which may not yet have been incorporated in the form of handbook revisions.

The pilot's flying experience is recognized; these instructions are not intended to teach the basic principles of flight, but are designed to provide the pilot with a general knowledge of the airplane.

This handbook is divided into sections as follows:

SECTION I, DESCRIPTION. This section describes the airplane and its systems and controls which contribute to the physical act of flying the airplane. Also included is emergency equipment which is not part of an emergency system.

SECTION II, NORMAL PROCEDURES. Included in this section are the procedures to be followed on a non-tactical flight under normal conditions, from the time the pilot approaches the airplane until the flight is completed and the airplane is left parked on the ramp.

SECTION III, EMERGENCY PROCEDURES. The procedures to be followed in meeting any emergency that the pilot could reasonably expect to encounter, except those in connection with auxiliary equipment, are described in this section.

SECTION IV, DESCRIPTION AND OPERATION OF AUXILIARY EQUIPMENT. This section includes description, normal operation and emergency operation of all equipment not directly contributing to flight but which enables the airplane to perform specialized functions. Portions of this section are incorporated in the supplement to this handbook, CO 01-85FGD-1A.

SECTION V, OPERATING LIMITATIONS. All important limitations which must be observed during normal operation of the airplane are covered in this section. This section is incorporated in the supplement to this handbook, CO 01-85FGD-1A.

SECTION VI, FLIGHT CHARACTERISTICS. This section describes the unique characteristics of this airplane in flight. This section is incorporated in the supplement to this handbook, CO 01-85FGD-1A.

SECTION VII, SYSTEMS OPERATION. Operation of the various airplane systems under varying conditions is discussed in this section, with emphasis given to any special problems which must be considered.

SECTION VIII, CREW DUTIES. Not applicable.

SECTION IX, ALL WEATHER OPERATION. This section contains procedures to be followed under extreme climatic conditions. No repetition is made of procedures given in other sections.

APPENDIX I, OPERATING DATA. This section contains all operating data charts and tables and explanatory text required for preflight and inflight mission planning. This section is incorporated in the supplement to this handbook, CO 01-85FGD-1A. Take-off data, however, is also incorporated in Appendix I of this handbook, AN 01-85FGD-1.

SECTION I

TABLE OF CONTENTS

SECTION I DESCRIPTION

AIRPLANE—F9F-6.

The airplane is a single place, single jet engine fighter, with swept-back wings and horizontal stabilizer, built by the Grumman Aircraft Engineering Corporation. It is designed primarily for aircraft carrier operations and is equipped with folding wings and an arresting gear. Rudder control on this airplane is conventional. On early airplanes, longitudinal control is accomplished by means of a conventional spring tab elevator control system. On later airplanes, longitudinal control is attained by a hydraulic powered, all movable horizontal stabilizer (flying tail). Lateral control is accomplished by hydraulically operated flaperons.

Principal dimensions are:

Span	34 ft 6 in.
Span (wings folded)	14 ft 2 in.
Length	40 ft 10 in.
Height	12 ft 1-1/2 in.
Height (wings folded)	15 ft 10 in.

AIRPLANE—F9F-6P.

This airplane is a modification of the F9F-6 to a photographic reconnaissance airplane. The modification to eliminate the guns and related equipment and incorporate the photographic equipment and automatic pilot and their controls and instruments has resulted in the following changes:

a. Rearrangement of electronics equipment installed in the area enclosed by the fuselage nose section, lengthening of this section by 12 inches, and shortening of the sliding nose section.

b. Rearrangement of the left and right consoles and the main instrument panel to provide space for the controls associated with the additional equipment.

c. Some minor changes of the fuselage structure and equipment installations to provide for the necessary ducting control for hot air from the engine compressor, which is used for defrosting the camera windows and heating the camera compartment.

d. Removal of all armament and the Armament Control System, removal of AN/APG-30 system and installation of an additional armor plate bulkhead.

ENGINE.

The J48-P-6A[1] or J48-P-8[2] engine is of the centrifugal flow, turbo-jet type. It is equipped with a fuel boost pump, two main fuel pumps and a fuel control unit. The fuel control unit automatically adjusts itself with engine speed and altitude changes.

ENGINE FUEL CONTROL SYSTEM.

The engine fuel system consists of an engine driven fuel boost pump, a low pressure filter, two engine driven

[1]Airplanes ser No. 126257 through 127326.
[2]Airplanes ser No. 127327 and subsequent.

Figure 1-1. F9F-6, -6P Airplanes

high pressure fuel pumps, a Holley fuel control unit, and a fuel pressurizing and shut-off valve. The Holley fuel control unit provides rapid, controlled acceleration within tailpipe temperature limitations, and also limits deceleration rates to prevent flame-out. The control includes temperature and pressure units designed to maintain proper rate of fuel flow during changes in airplane speed and altitude. Transfer from the primary (normal operating) fuel system to the emergency fuel system in case of failure of the primary system is accomplished by setting the engine fuel system selector switch to its emergency position. A red warning light is installed to warn the pilot of failure of either of the main pumps.

NORMAL OPERATION.

In normal operation, fuel from the front tank flows through the engine driven booster pump and filter to the main pumps. A common tube joining these pumps leads fuel to the control unit. Fuel leaves the control unit through one side of a double ball check valve and is routed through a pressurizing and shut-off valve to the combustion chambers. At low engine speed, this valve supplies fuel to the central swirl passage in each nozzle through a pilot manifold. As engine speed and consequent pump pressure increase, the pressurizing valve opens progressively, until at maximum rpm, the valve is fully open and the nozzles receive the bulk of their fuel supply from the main fuel manifold. The shut-off valve is opened by the throttle linkage when the latter is moved from START to IDLE.

EMERGENCY OPERATION.

Operation of the emergency fuel system is initiated by the pilot setting the engine fuel system selector switch to its emergency position. When a changeover to emergency is made or when the fuel pressure in the primary system drops below the actuating pressure, the amber warning light on the fuel control panel will glow; a normally open valve in the primary system will close; while a normally closed valve in the emergency system will open, directing fuel through the emergency control. When operating on the emergency system, there is no automatic acceleration or deceleration control or temperature or density compensation, and the pilot must operate the throttle carefully to avoid overspeeding or overheating the engine. During emergency operation, the engine high pressure pumps are isolated from the fuel control hydraulically and are therefore at their maximum pressure output.

At sea level and low altitudes, the emergency control may be selected and obtained safely with the throttle in any operating position. If the primary fuel control malfunctions during take-off, the emergency fuel control may be selected IMMEDIATELY without having to retard the throttle to IDLE.

Note

If switching to the emergency fuel control is delayed until rpm has fallen off to less than 80%, it would be advisable, time permitting, to first retard the throttle to IDLE, then to switch to the emergency fuel control and advance the throttle smoothly to obtain the thrust needed for flight.

ENGINE FUEL SYSTEM CONTROLS.

FUEL MASTER SWITCH.

This toggle switch (44, figure 1-3B), equipped with a guard, is located on the right console electrical control panel, outboard. When set to ON, it energizes the submerged low pressure fuel boost pump in the bottom of the front tank and opens the valve in the feed line to pump fuel to the engine. When the switch is set to OFF, the valve is closed and pump operation is stopped.

CAUTION

This switch MUST BE ON whenever the engine is running. If the switch is set at OFF, fuel feed will be shut off, possibly causing line collapse and fuel line trouble and, since the engine fuel pumps have no dry rating, eventual pump failure. This switch must not be set to OFF until the throttle is moved to CLOSED and the engine has stopped rotating.

ENGINE FUEL SYSTEM SELECTOR SWITCH—J48-P-6A ENGINE.

This three position toggle switch (33, figure 1-2A) is on the fuel control panel on the left console. The three positions are forward (momentary)—RESET PRIMARY, center — PRIMARY, and aft — START & EMER. When set to START & EMER. for ground or air starting, or in case of failure of the primary system, the engine fuel control system is transferred to operation on the emergency system from the primary system. In this condition all fuel control is manually regulated with the throttle. The amber warning light on the panel glows with this switch setting. When the switch is set to RESET PRIMARY and held momentarily before being set to PRIMARY, the fuel control system is reset to normal operation (on the primary system) after operating on the emergency system.

CAUTION

This switch should be set at START & EMER. only for starting or for actual emergency conditions, as this setting places very severe requirements on the main pumps.

ENGINE FUEL SYSTEM SELECTOR SWITCH—J48-P-8 ENGINE.

This three position toggle switch (33, figure 1-2A) is on the fuel control panel on the left console. The three positions are forward (momentary)—RESET PRIMARY, center—START & PRIMARY, and aft—EMER. Plac-

1. AN/APG-30 Armament Control System Radar Antenna (See Note 1)
2. 20 MM Guns (4)
3. AN/ARN-6 Radio Compass Loop Antenna
4. Inboard Guns Ammunition Boxes (2)
5. Battery
6. Right Cabin Electrical Distribution Box
7. IFF Detonator Impact Switch Box
8. Brake System Hydraulic Reservoir
9. Electrical System Circuit Breaker Panel
10. Gun Camera
11. ID-295/APG-30 Armament Control Tracking Indicator
12. Armament Control Gun Sight
13. Electronic System Circuit Breaker Panel
14. Ejection Seat Face Cover Handles and Head Rest
15. Air Conditioning Turbine Oil Fillerneck and Drain
16. Oxygen Bottles
17. AN/ARN-6 Radio Compass Sense Antenna
18. Canopy Unlatch (for ground operation)
19. Forward Fuel Tank Fillerneck
20. Fuselage Light
21. Aft Tank Fillerneck
22. Hydraulic System Accumulator
23. Upper Hydraulic Reservoir and Fillerneck
24. Generator Electrical Connector Box
25. Engine Electrical Connector Box
26. Engine Oil Fillerneck
27. Engine Driven Accessories
28. J48-P-6A, -8 Engine
29. AN/ARC-1 VHF or AN/ARC-27 UHF Command Set Antenna
30. Tail Position Lights
31. Arresting Hook Recoil Strut
32. Tail Skid
33. Formation Light L/R
34. Wing Tank Dump Valve L/R
35. Running Light L/R
36. D-C External Power Receptacle
37. Wing Tank Fuel Cells
38. Lower Hydraulic Reservoir
39. Fuel Tank Water Drain Valve
40. Pressure Fueling Valve Wing Tanks
41. Wing Tank Fillerneck L/R
42. AN/APN-1 Radio Altimeter Antenna
43. Approach Light
44. Fuel System Drain Valve
45. Pressure Fueling Valve Fuselage Tanks
46. AN/APX-6 IFF Antenna
47. Landing Gear Emergency Air Bottle
48. Outboard Guns Ammunition Box
49. Boarding Ladder
50. Electric Auxiliary Hydraulic Pump

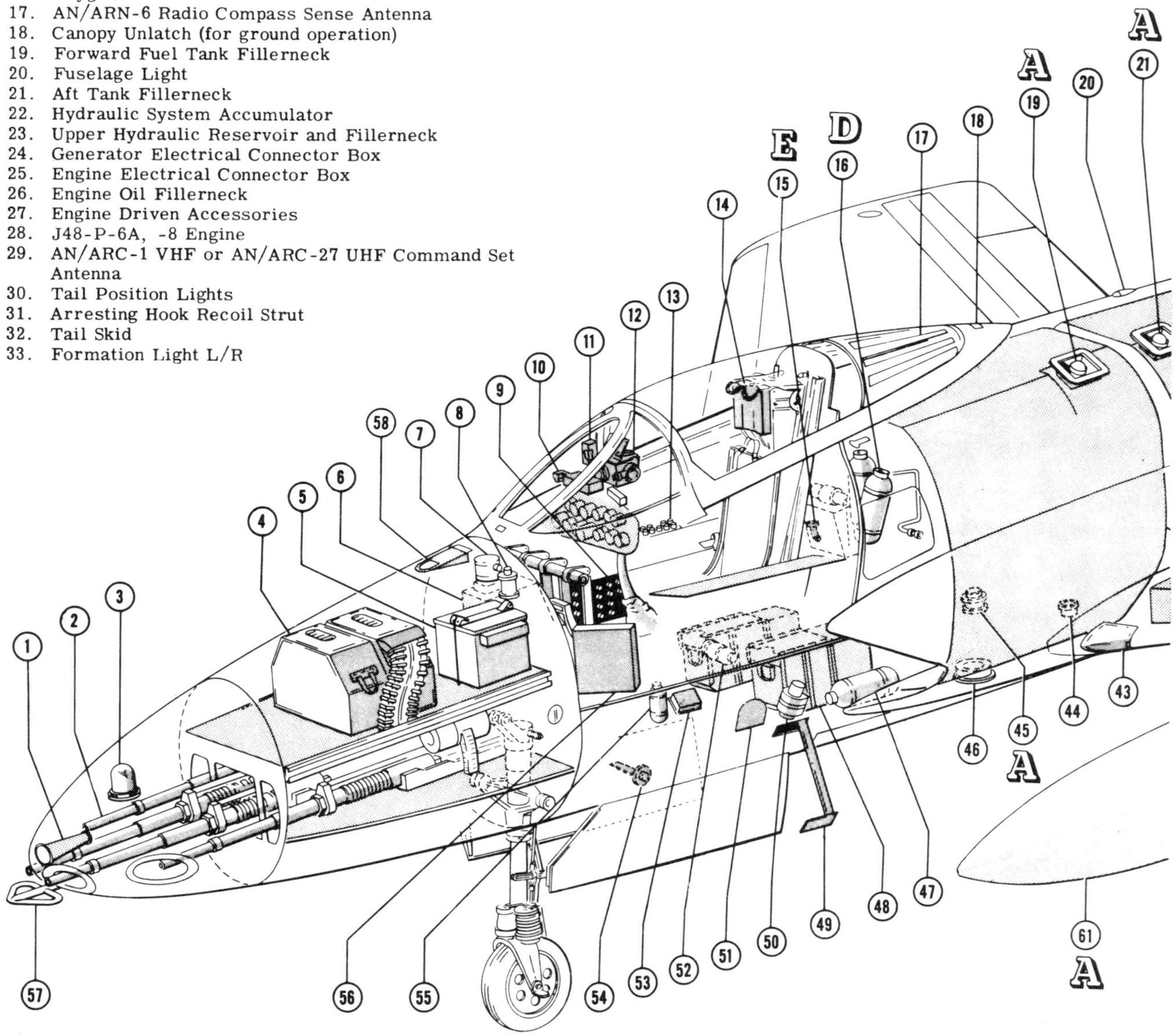

Figure 1-2. General Arrangement and Servicing (Sheet 1)

51. Step and Ladder Release
52. Canopy Emergency Air Bottle
53. A-C External Power Receptacle
54. AN/ARR-2A Navigational Receiver Antenna
55. Brake and Seat Ejection Emergency Air Bottle
56. Left Cabin Electrical Connector Box
57. Barricade Deflector
58. AN/APG-30 Armament Control System Radar Antenna (See Note 2)
59. Movable Stabilizer Actuator
60. Flaperette Hydraulic Reservoir & Accumulator
61. Droppable Tank (or External Stores) L/R

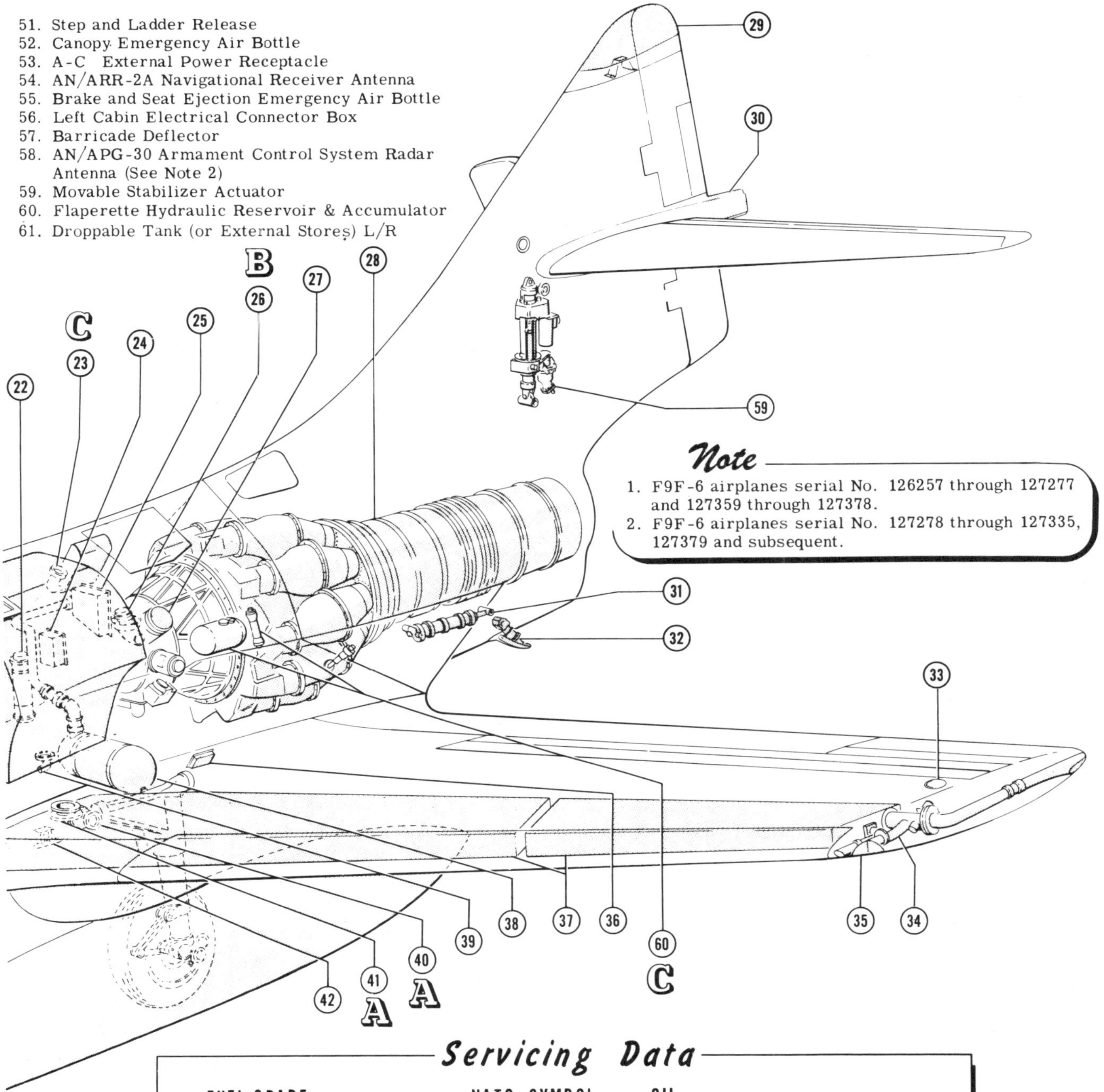

Note

1. F9F-6 airplanes serial No. 126257 through 127277 and 127359 through 127378.
2. F9F-6 airplanes serial No. 127278 through 127335, 127379 and subsequent.

Servicing Data

FUEL GRADE / **NATO SYMBOL**

			NATO Symbol
A	Recommended	JP-4(MIL-F-5624)	F-40
		JET-MIX	None
	Alternate	JP-3(MIL-F-5624)	None
		AVGAS 80(MIL-F-5572)	F-12
		AVGAS 91/96(MIL-F-5572)	F-15
		AVGAS 100/130(MIL-F-5572)	F-18
		AVGAS 115/145(MIL-F-5572)	F-22

Note

Appendix I operating data based on JP-3 or JP-4 are applicable to operation using JET-MIX. Maximum safe pressure and flow for pressure fueling - 200 gpm at 20 psi.

OIL

B Spec MIL-O-6081 Grade 1010
Tank Capacity 13 US Quarts

HYDRAULIC FLUID

C Spec MIL-O-5606 or MIL-F-7083 (Aer)
(as indicated on reservoir)
Main Reservoir Capacity 6 US Gal
Flaperette Reservoir Capacity 1.8 US Gal

OXYGEN

D Pressure 1800 ±50 psi

COOLING TURBINE OIL

E Spec MIL-O-6085
Capacity 50 cc

Figure 1-2. General Arrangement and Servicing (Sheet 2)

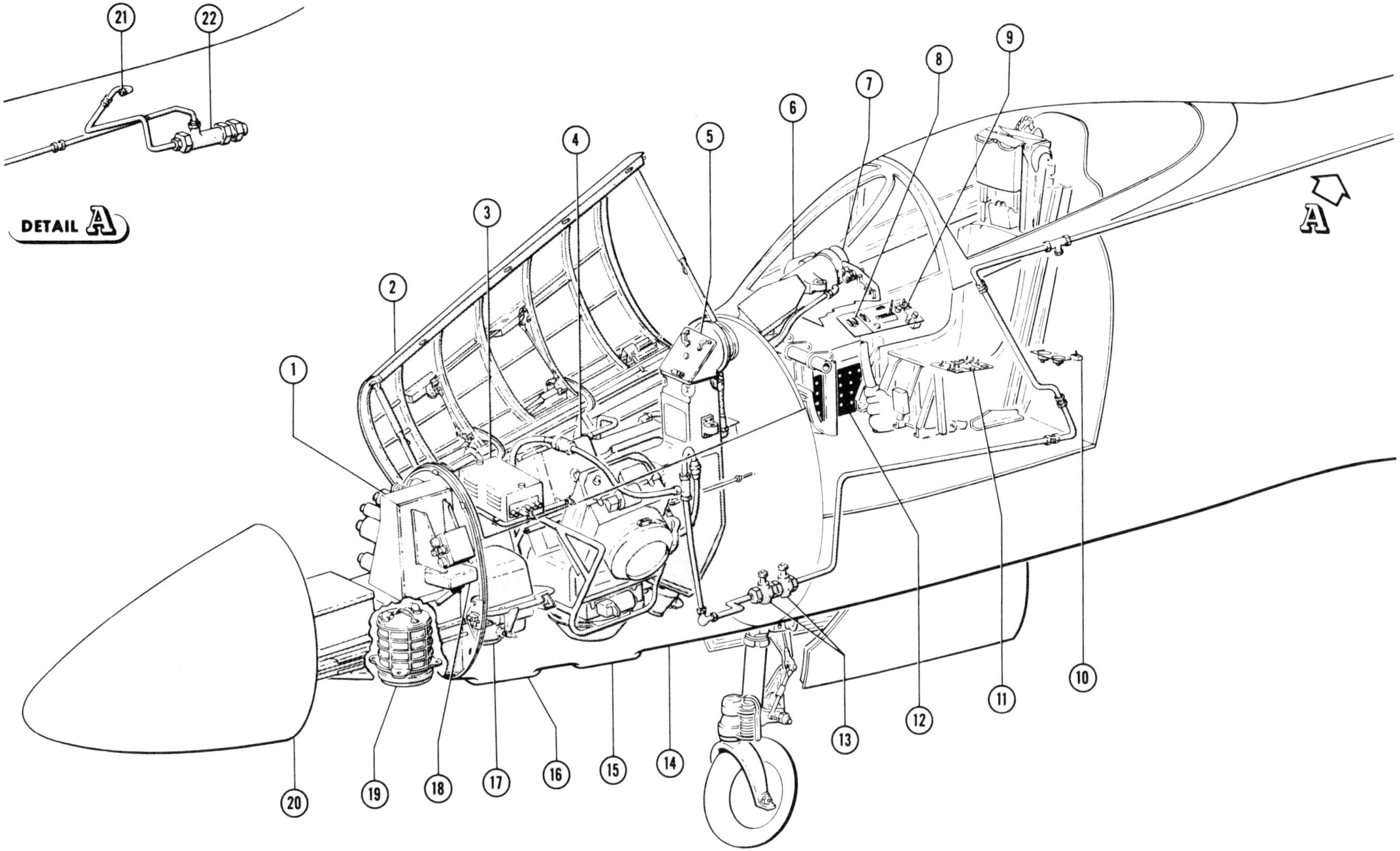

1. Photo Recorder
2. Camera Access Hatch
3. Sonne Amplifier
4. Trimetrogon Camera in Aft Bay
5. Viewfinder
6. Viewfinder Controls and Indicators
7. Viewfinder Eyepiece
8. Interval Computer (Multi-Camera Interval Control)
9. Right Console - Auto Pilot Controls
10. Camera Compartment Heat Control
11. Left Console - Camera Controls
12. Circuit Breaker Panel
13. Relief Valves
14. Viewer Compartment
15. Aft Camera Bay Window
16. Forward Camera Bay Window
17. K-17-6 in. Camera in Fwd Bay
18. Photo Recorder Gun Camera
19. Scanner
20. Sliding Nose Section
21. Connection to Engine Air at sta 258
22. Jet Air Ejector Pump

Figure 1-2. General Arrangement and Servicing (Sheet 3)

ing the switch to START & PRIMARY permits starting the engine and operating on the primary (automatic fuel control) fuel system.

Note

The amber emergency fuel system warning light will glow during the first part of the starting cycle until fuel pressure in the primary system builds up to required operating pressure.

Setting the switch to EMER. controls the changeover from the primary fuel system to the emergency fuel system.

Note

Ground or airstarts may be made with the switch set at EMER. as well as at START & PRIMARY.

When operating on the emergency system, the fuel control is manually regulated by the throttle. The amber warning light on the panel glows with this switch setting. When the switch is set to RESET PRIMARY and held momentarily before being set to START & PRIMARY, the fuel control system is reset to normal operation (on the primary system) after having been operating on the emergency system.

CAUTION

This switch should be set at EMER. only for actual emergency conditions, as this setting places very severe requirements on the main pumps.

FUEL SYSTEM BOOST PUMP CUT-OFF SWITCH BUTTON.

This switch push button is on the fuel control panel on the left console (32, figure 1-2A). With the engine operating, pressing this button cuts out the airplane fuel system boost pump in the main fuel tank, so that the engine driven boost pump is the only source of pressure to the two engine driven high pressure fuel pumps. If pressure from this pump is below that required, the low fuel boost pressure light will glow.

THROTTLE.

The throttle lever is installed on the quadrant on the left console (22, figure 1-2A). When it is set full aft and inboard at the CLOSED position, the engine is shut down, the high pressure cock is closed and the ignition is off. When the lever is moved outboard to the START position, it actuates a switch to supply power to the spark igniters with ignition on (the engine start master switch must have been set to ON and the engine cranking switch set to START). When it is moved forward toward the IDLE position, its linkage mechanically opens the high pressure cock, and when moved inboard past the detent and into the IDLE position, fuel is fed to the engine to operate it at the proper idle speed. When the throttle is moved to its full open position, engine should operate at 100% (11000 rpm).

THROTTLE FRICTION CONTROL.

A throttle friction control lever is installed on the inboard side of the quadrant (19, figure 1-2A). The lever is moved forward to increase friction on throttle and aft to decrease.

CATAPULT GRIP.

A catapult grip is installed just forward of the throttle quadrant (18, figure 1-2A). The grip is a spring loaded rod, normally stowed. For catapult take-off use, the tab on the top is rotated inboard, which permits pulling the grip up into position. The grip is rotated outboard and drops down to the stowed position when its use is not desired.

ENGINE FUEL SYSTEM INDICATORS.

ENGINE FUEL PUMPS WARNING LIGHT.

A red warning light for the two engine driven (high pressure) fuel pumps is on the fuel control panel on the left console (34, figure 1-2A). The light glows to indicate failure of either pump.

EMERGENCY FUEL SYSTEM INDICATOR LIGHT.

This amber warning light (31, figure 1-2A or 36, figure 1-3) glows when the fuel pressure in the primary fuel system is below normal and, therefore, will be on when the engine is operating on the emergency fuel system. It is located on the fuel control panel on the left console.

LOW FUEL BOOST PRESSURE WARNING LIGHT.

A red warning light on the upper left corner of the main instrument panel (7, figure 1-3A) glows when pressure from the airplane fuel system boost pump drops below 6 psi or when the fuel filter is blocked by ice or foreign matter. The latter condition indicates that the engine is operating on unfiltered fuel. Prior to engine starting, operation of the airplane fuel system (fuselage tank) boost pump may be checked by setting the fuel master switch to ON and observing that the low fuel boost pressure warning light glows momentarily, an indication that the pump has built up the required pressure. While the engine is running, the engine driven boost pump may be checked by depressing the fuel boost pump cut-off switch button and observing that the low fuel boost pressure warning light does not glow, an indication that this pump is holding the required pressure above 6 psi. "Light out" condition on both of these checks also indicates that the filter is operating normally and is unclogged.

Note

The use of a single light for the dual purpose described does not permit identification of malfunction. It warns the pilot only of trouble in the low pressure portion of the fuel system.

ENGINE COOLING.

Engine cooling is accomplished by two separate automatic systems. The first system consists of a cooling air impeller mounted on the compressor shaft which receives

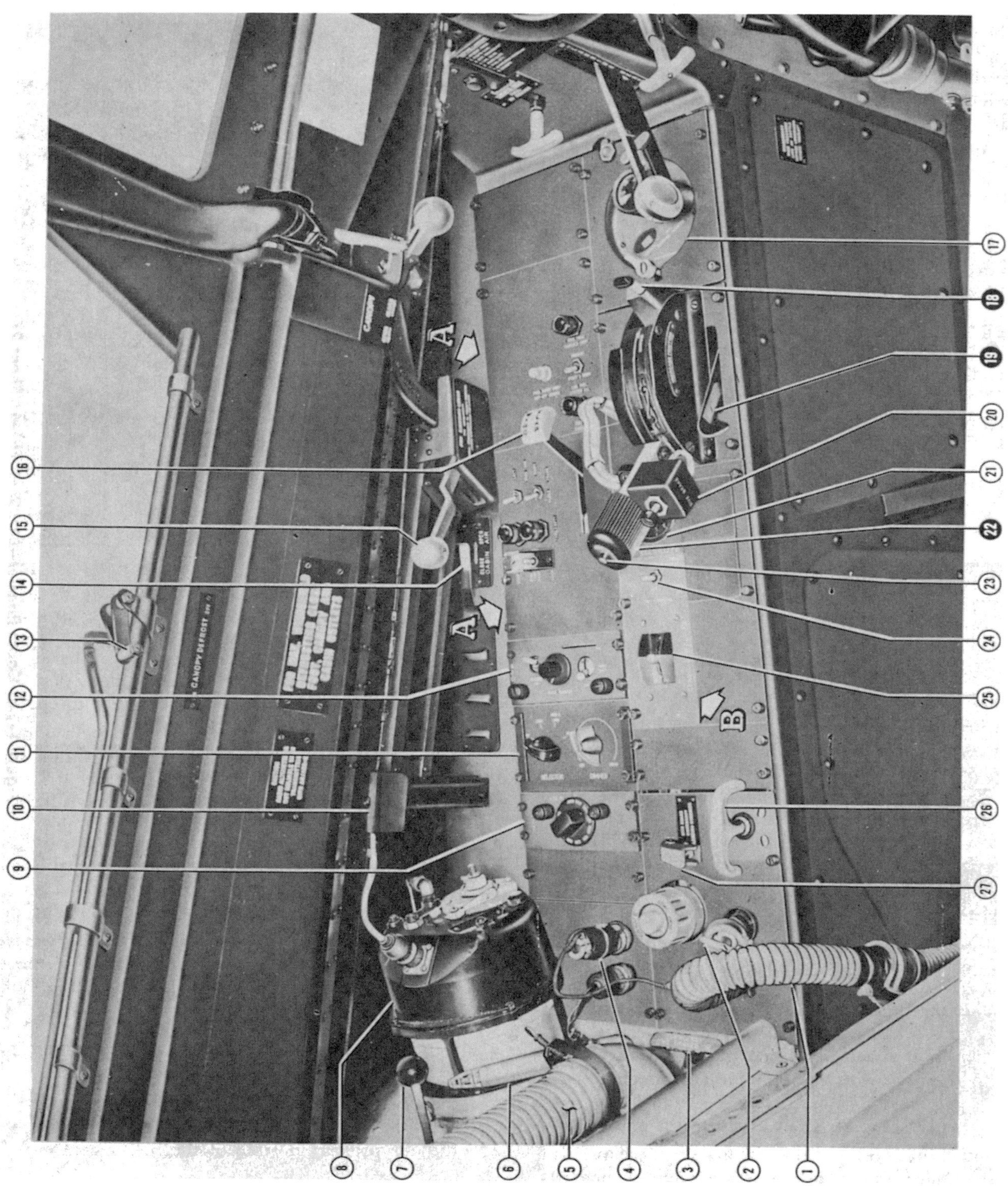

Figure 1-2A. Cabin Interior—Left Side (F9F-6) (Sheet 1)

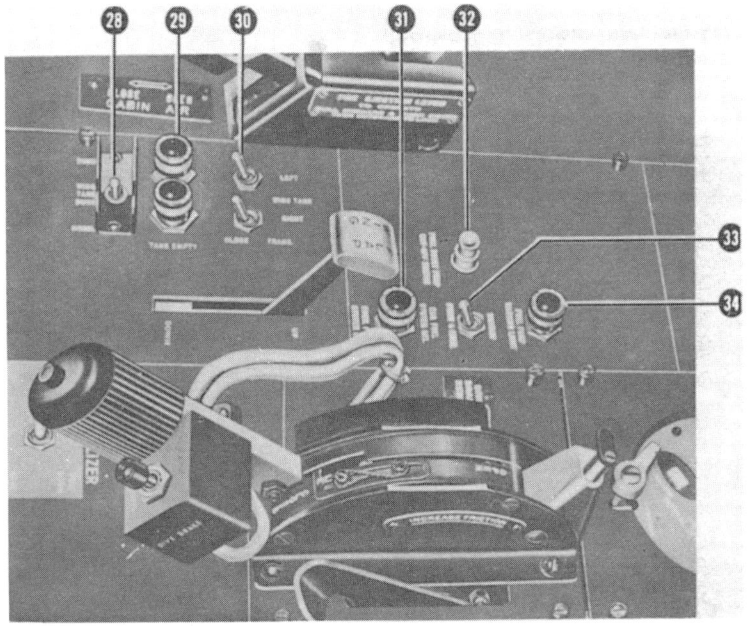

DETAIL A

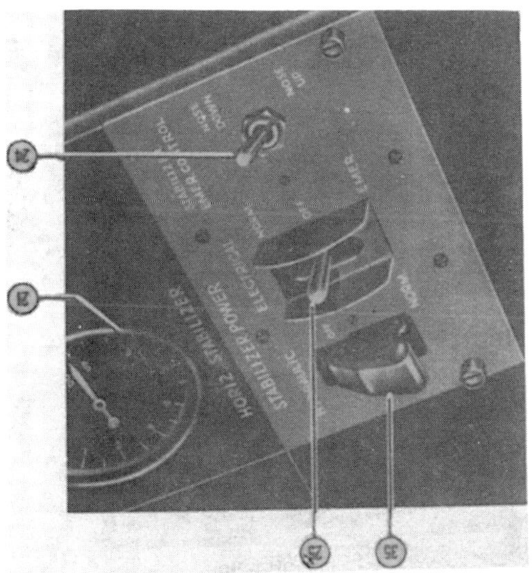

DETAIL B

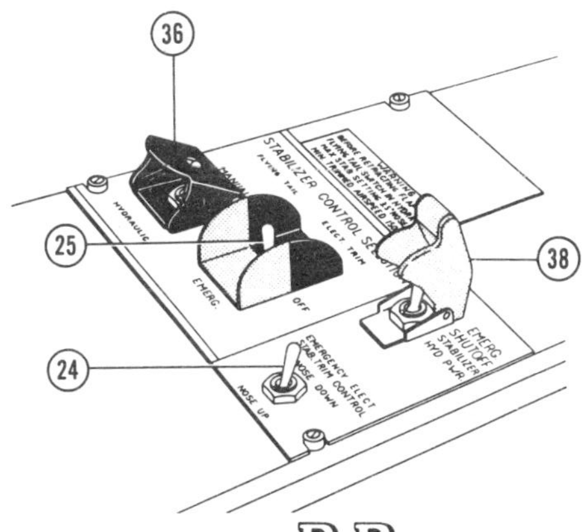

DETAIL BB

1. Oxygen Tube (Personal Gear Composite Disconnect to Mask)
2. Anti-"G" Suit Tube Receptacle (Personal Gear Composite Disconnect to Suit)
3. Emergency Speed Brakes Up Control Handle (Override Control)
4. Microphone and Headset Plug (Personal Gear Composite Disconnect to Microphone and Headset)
5. Canopy Defroster Tube
6. Cabin Air Conditioning Thermostat
7. Cabin Pressure Dump Control Lever
8. Cabin Pressure Regulator
9. Radio Altimeter Limit Switch (See Note 1)
10. Tab for Personal Gear Connections
11. MK 20 Mod 0 A. C. S. Control Rheostats
12. C-775/APG-30 Gun Ranging Radar Control
13. Canopy Defroster Control Lever
14. Console Air Conditioning Outlet Control Handle
15. Ejection Seat Pre-ejection Lever
16. Wing Flap Control Lever
17. Oxygen Regulator and Control Panel
18. Catapult Grip
19. Throttle Friction Control Lever
20. Speed Brakes Control Switch
21. Stabilizer Position Indicator
22. Throttle Lever and Aero 4C Radar Ranging Control (Throttle Grip)
23. Microphone Button Switch
24. Stabilizer Trim Emergency Control Switch
25. Stabilizer Trim Emergency Power Control Switch
26. Brake Emergency Control Handle
27. Speed Brakes Emergency Landing Override Switch
28. Wing Tank Dump Switch
29. Wing Tank Empty Warning Lights
30. Wing Tank Selector Switches
31. Emergency Engine Fuel System Warning Light
32. Fuel System Boost Pump Cut-Off Button Switch
33. Engine Fuel System Selector Switch
34. Engine Driven Fuel Pumps Warning Light
35. Hydraulic (Emergency Stabilizer Shift) Control Switch (See Note 2)
36. Longitudinal Control System Selector Switch
37. DELETED
38. Stabilizer Hydraulic Power Emergency Shut-Off Switch

Note

1. Radio Altimeter Limit Setting Switch is installed only on airplanes ser No. 126257 through 127297.
2. On airplanes ser No. 127258 and subsequent, the stabilizer power control panel (Detail B) replaces stabilizer power control panel shown in main view. Item 35 is operative only when the flying tail system is installed.
3. On airplanes with the interim flying tail modification installed, the panel shown as detail BB replaces the one shown as detail B. On these airplanes the stabilizer position indicator, item 21, is replaced by an absolute stabilizer position indicator mounted on the instrument panel.

Figure 1-2A. Cabin Interior—Left Side (F9F-6) (Sheet 2)

Figure 1-3. Cabin Interior—Left Side (F9F-6) (Sheet 1)

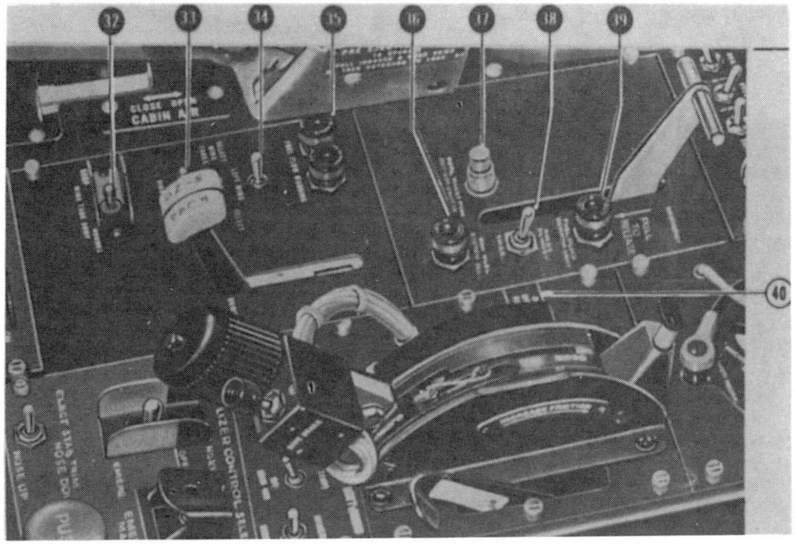

DETAIL A

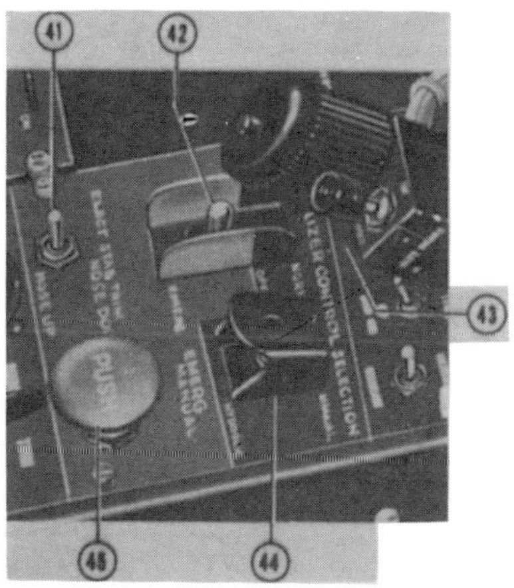

DETAIL B

1. Anti - "G" Suit Tube Receptacle (Personal Gear Composite Disconnect to Suit)
2. Emergency Speed Brakes Up Control Handle (Override Control)
3. Anti - "G" Suit Valve Control Knob
4. Microphone and Headset Plug (Personal Gear Composite Disconnect to Microphone and Headset)
5. Cabin Pressure Regulator
6. Cabin Pressure Dump Control Lever
7. Emergency Flaperette Power Air Pressure Gage (See Note 1)
8. Oxygen Tube (Personal Gear Composite Disconnect to Mask)
9. Canopy Defroster Control Lever
10. Mk 20 Mod 0 A.C.S. Control Rheostats
11. Gun Ranging Radar Control Panel
12. Console Air Conditioning Outlet Control Handle
13. Pre-ejection Lever
14. Wing Flap Control Lever
15. Canopy Emergency (Air) Control Lever
16. Canopy Normal (Hydraulic) Control Lever
17. In-flight Refueling Switch
18. Armament Control Panel (See Note 2)
19. Eternal Stores Emergency Jettison Tee Handle Control
20. Oxygen Regulator and Control Panel
21. Catapult Grip
22. Throttle Friction Control Lever
23. Air Conditioning Control Panel (See Note 4)
24. Speed Brakes Control Switch
25. Throttle Lever and Aero 4C Radar Ranging Control (Throttle Grip)
26. Microphone Switch
27. Rudder Trim Control Switch
28. Rudder Trim Position Indicator
29. Emergency Flaperette Power Control Lever (See Note 1)
30. Wheel Brakes Emergency Control Handle
31. Speed Brakes Emergency Landing Override Switch
32. Wing Tank Dump Switch
33. Wing and Droppable Fuel Tank Selector Switch, (Left Wing) (See Note 3)
34. Wing and Droppable Fuel Tank Selector Switch (Right Wing) (See Note 3)
35. Fuel Flow Warning Lights (Left and Right Wing Tanks) (See Note 3)
36. Emergency Engine Fuel System Indicator Light
37. Fuel System Boost Pump Cut-off Switch
38. Engine Fuel System Selector Switch
39. Engine Driven Fuel Pumps Warning Light
40. Speed Brakes Circuit Breaker
41. Emergency Trim Control Switch (See Note 5)
42. Electrical Trim Selection Switch (See Note 5)
43. Flying Tail Circuit Breaker (behind throttle) (See Note 5)
44. Longitudinal Control System Selector Switch (See Note 5)
45. Emergency Manual Control Knob (See Note 5)

Note

1. Emergency flaperette control lever and air gage installed on airplanes with J48-P-8 engines.
2. The armament control panel is installed on airplanes ser No. 128155 and subsequent.
3. Wing and droppable fuel tank selector switches and fuel flow warning lights installed on airplanes with external stores provisions.
4. The air conditioning control panel is installed on airplanes ser No. 128231 and subsequent.
5. The stabilizer control selection panel shown as detail B is installed on F9F-6 airplanes ser No. 127336 through 127358, 131037 and subsequent. It will be installed on all airplanes when F9F Aircraft Service Change No. 256 has been incorporated.

Figure 1-3. Cabin Interior—Left Side (F9F-6) (Sheet 2)

Figure 1-3A. Cabin Interior—Forward (F9F-6)

KEY TO FIGURE 1-3A

1. Landing Gear Emergency (Air) Control Handle
2. Wheels and Flaps Position Indicator
3. Canopy Normal (Hydraulic) Control Lever
4. Canopy Emergency (Air) Control Lever
5. Radio Altimeter Low Limit Warning Light (See Note)
6. Oil Pressure Indicator
7. Low Fuel Boost Pressure Warning Light
8. Speed Brakes Position Indicator
9. Radio Altimeter (See Note)
10. Armament Switch Panel
11. Sight Unit
12. Tracking Indicator, ID-295/APG-30
13. Stand-by Compass
14. Elapsed Time Clock
15. Accelerometer
16. Fire Warning Lights
17. Fuel Quantity Indicator, Push-to-Test Switch
18. Fire Warning Light Circuit Test Switch
19. Low Fuel Level Warning Light
20. Clock
21. Fuel Flowmeter
22. Hook Position Warning Light
23. Tail Skid Control Switch
24. Cabin Altimeter
25. Sliding Nose Unlocked Warning Light
26. Arresting Hook Control Handle
27. Hook Up Push Button Switch
28. Fuel Quantity Indicator
29. Cabin Air Foot Outlet Control
30. Radio Compass Indicator
31. Fast Erecting Gyro Horizon
32. Rate-of-Climb Indicator
33. G-2 Remote Compass
34. Turn and Bank Indicator
35. Wing Trimmer and Stabilizer Trim Thumb Switch
36. Relief Tube
37. Radar Switch
38. Bomb Release Switch
39. Altimeter
40. Maximum Allowable Airspeed Indicator
41. Tailpipe Temperature Gage
42. Landing Gear Control, Down Lock Solenoid, Manual Release Knob
43. Airstart Emergency Igniter System Control Handle
44. Landing Gear Normal Control Lever and Landing Gear Unlocked Warning Light
45. Tachometer

NOTE

Radio altimeter and low limit warning light are removed by Service Change.

On F9F-6 airplanes with the interim flying tail modification installed, an absolute stabilizer position indicator is installed on the upper left side of the instrument panel.

Figure 1-3A-1. Cabin Interior—Forward (F9F-6)

KEY TO FIGURE 1-3A-1

1. Landing Gear Emergency (Air) Control Handle
2. Tailpipe Temperature Indicator
3. Wheels and Flaps Position Indicator
4. Tachometer
5. Low Fuel Boost Pressure Warning Light
6. Oil Pressure Indicator
7. Speed Brakes Position Indicator
8. Accelerometer
9. Armament Master Switch
10. Inboard and Outboard Gun Charging Switches
11. Stores Jettison Switch
12. Location of Absolute Stabilizer Position Indicator
13. Sight Unit Range Scale
14. Sight Unit
15. Sight Unit Target Span Handle
16. Tracking Indicator
17. Location of Stand-by Compass
18. Clock
19. Location of Elapsed Time Clock
20. Fire Warning Lights
21. Fire Warning Light Circuit Test Switch
22. Fuel Quantity Indicator Push-to-Test Switch
23. Low Fuel Level Warning Light
24. Fuel Flowmeter
25. Fuel Quantity Indicator
26. Cabin Pressure Altitmeter
27. Arresting Hook and Barrier Guard Control Handle
28. Arresting Hook Raising Push Button Switch
29. Air Conditioning Foot Outlet Control
30. Course Indicator
31. Gyro Horizon Indicator
32. Rate of Climb Indicator
33. Turn and Bank Indicator
34. G-2 Remote Compass Indicator
35. Altimeter
36. Maximum Allowable Airspeed Indicator
37. Airstart Emergency Igniter Control
38. Landing Gear Control, Down Lock Solenoid Manual Release Knob
39. Landing Gear Normal Control Lever and Landing Gear Unlocked Warning Light

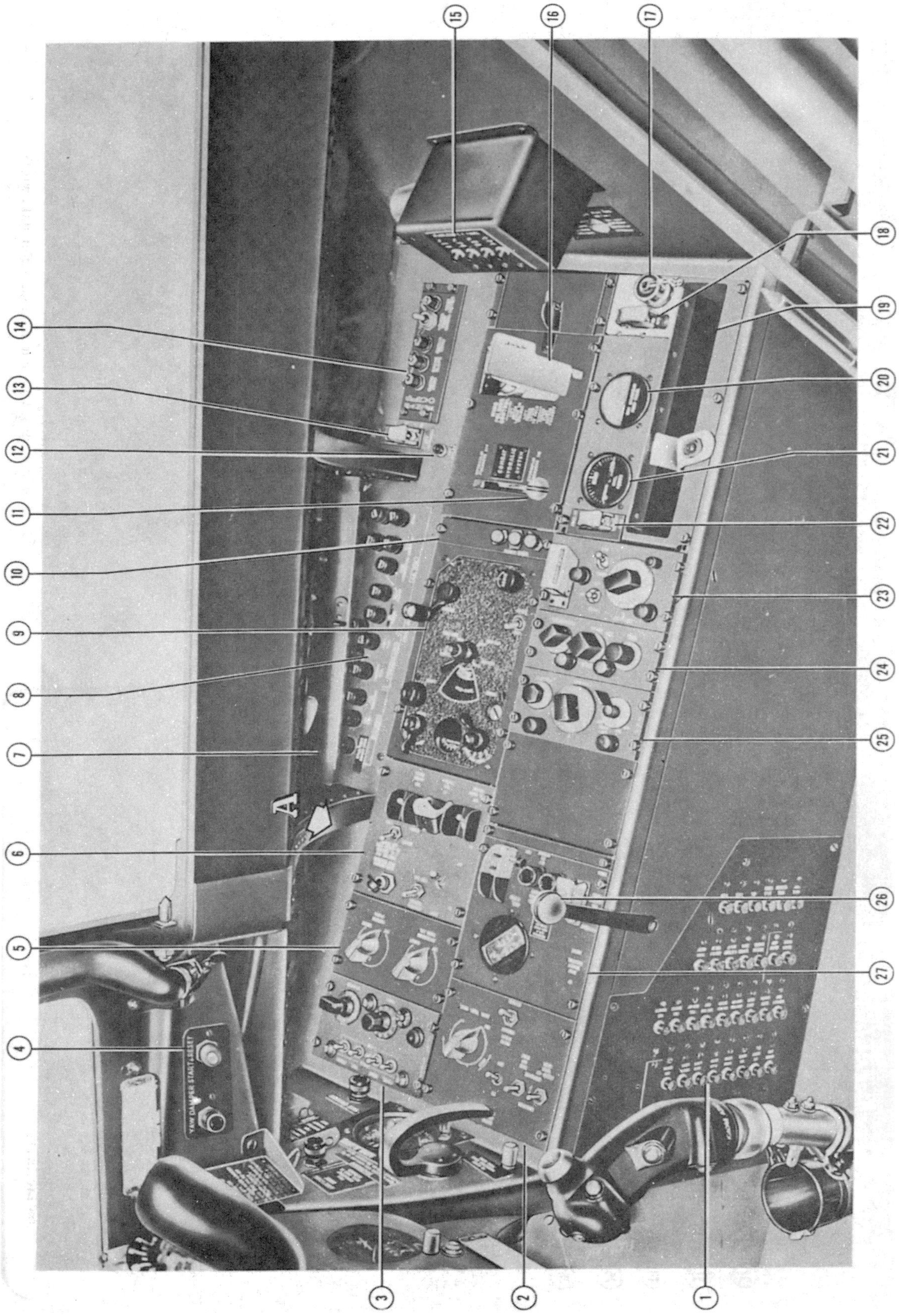

Figure 1-3B. Cabin Interior—Right Side (F9F-6) (Sheet 1)

DETAIL A

Note

1. Installed in airplanes ser No. 127258 and subsequent.
2. Antenna Relay Switch is operative only when AN/ARC-27 UHF Command Set is installed.
3. AN/ARC-27 UHF Command Set may be replaced by the AN/ARC-1 VHF Command Set depending upon local communication facilities.

1. Circuit Breaker Panel
2. Air Conditioning System Control Panel
3. Exterior Lights Control Panel
4. Yaw Damper Start and Reset Switch and Warning Light Panel (See Note 1)
5. Interior Lights Control Panel
6. Engine and Miscellaneous Control Switches
7. Chartboard Stowage Case
8. Fuse Panel
9. AN/ARN-6 Radio Compass Control Panel
10. Spare Lamps
11. Hydraulic Pressure On-Off Control Lever
12. Antenna Relay Circuit Breaker (See Note 2)
13. Antenna Relay Control Switch (See Note 2)
14. Unit II, Servo Radar Range Aero 1A
15. Stabilizer Trim Power Circuit Breaker Panel
16. Control modified, see figure 1-10
17. Utility Receptacle
18. Auxiliary Hydraulic Pump Switch
19. Map Case
20. Auxiliary Hydraulic Pump Pressure Gage
21. Main Hydraulic System Pressure Gage
22. Flaperette System Hydraulic Pump Control Switch
23. AN/APX-6 IFF Control Panel
24. AN/ARR-2 Navigational Radio Control Panel
25. AN/ARC-27 UHF Command Set Control Panel (See Note 3)
26. Seat Height Control Lever
27. Electrical System Control Panel (See Detail A)
28. Stand-by Temperature Control Switch
29. Ram Air Ventilation Control Switch
30. Air Conditioning System On-Off Switch
31. Automatic Temperature Control Rheostat
32. Manual Exterior Lights Coding Indicator Light
33. Manual Exterior Lights Coding Key
34. Exterior Lights Control Switches (4)
35. Automatic Coding Letter Selector Switch
36. Exterior Lights Master Control Switch
37. Instrument Panel Lights Control Switch and Rheostat
38. Console Light Control Switch and Rheostat
39. G-2 Compass Control Switch
40. Fast Erecting Gyro Horizon Warning Light
41. Pitot Heat Control Switch
42. Fast Erecting Gyro Horizon Control Switch
43. Engine Cranking Switch
44. Fuel System Master Switch
45. Engine Starting Master Switch
46. Battery Switch
47. Generator Warning Light
48. Inverter Warning Light
49. Inverter Changeover Switch
50. Voltammeter
51. Cabin Pressurization Combat - Normal Switch

Figure 1-3B. Cabin Interior—Right Side (F9F-6) (Sheet 2)

Figure 1-3B-1. Cabin Interior—Right Side (F9F-6) (Sheet 1)

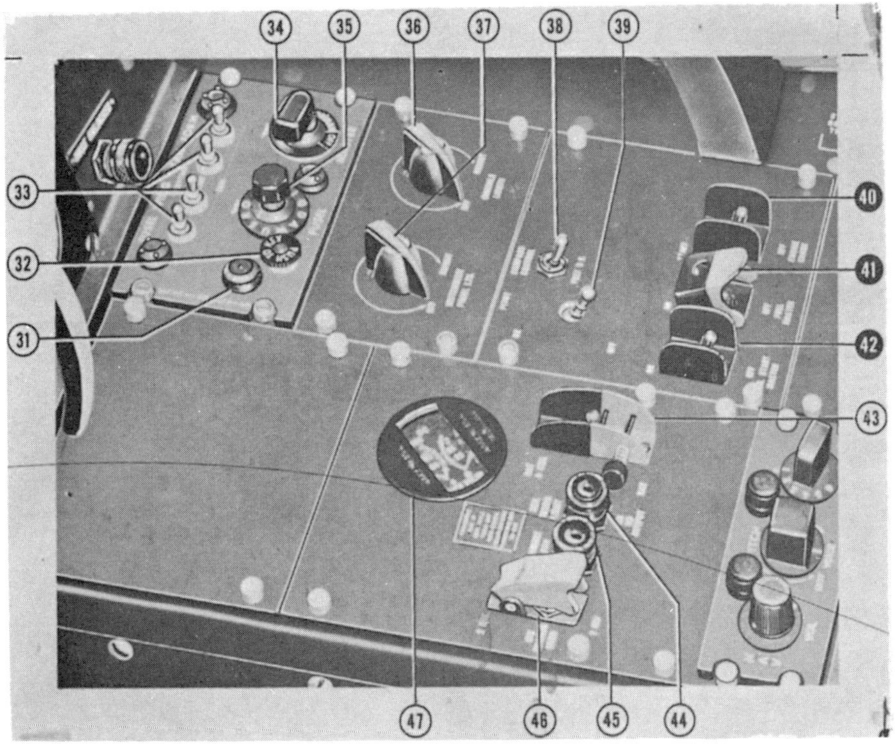

1. Circuit Breaker Panel
2. Exterior Lights Control Panel
3. Sliding Nose Unlocked Warning Light
4. Arresting Hook Position Warning Light
5. Tail Skid Control Switch
6. Yaw Damper Sensitivity Selector Switch (See Note)
7. Yaw Damper Power Switch (See Note)
8. Interior Lights Control Panel
9. Engine and Miscellaneous Control Switches
10. Fuse Panel
11. AN/APX-6 IFF Control Panel
12. Chartboard Stowage Case
13. AN/ARC-27 UHF Command Set Control Panel
14. Spare Lamp Stowage Panel
15. Antenna Relay Switch Circuit Breaker
16. Antenna Relay Control Switch
17. Unit II, Servo Radar Range Aero 1A
18. Stabilizer Control Power Circuit Breaker Panel
19. Control modified, see figure 1-10
20. Spare Lamp and Fuse Container
21. Utility Receptacle
22. Auxiliary Hydraulic Pump Control Switch
23. Map Case
24. Auxiliary Hydraulic Pump Pressure Gage
25. Main Hydraulic Pump Pressure Gage
26. Hydraulic Pressure On-Off Control Lever
27. Flaperette System Control Switch
28. AN/ARN-6 Radio Compass Control Panel
29. AN/ARR-2 Navigational Radio Control Panel
30. Electrical System Control Panel
31. Manual Exterior Lights Coding Indicator Light
32. Manual Exterior Lights Coding Key
33. Exterior Lights Control Switches (4)
34. Exterior Lights Master Control Switch
35. Automatic Coding Letter Selector Switch
36. Console Lights Control Switch and Rheostat
37. Instrument Panel Lights Control Switch and Rheostat
38. G-2 Compass Control Switch
39. Pitot Heat Control Switch
40. Engine Cranking Switch
41. Fuel Master Switch
42. Engine Start Master Switch
43. Battery Switch
44. Generator Warning Light
45. Inverter Warning Light
46. Inverter Changeover Switch
47. Voltammeter

Note

The electronic yaw damper system is installed on F9F-6 airplanes serial No. 130794 and subsequent.

Figure 1-3B-1. Cabin Interior—Right Side (F9F-6) (Sheet 2)

Figure 1-3C. Cabin Interior—Left Side (F9F-6P Airplanes ser No. 127473 through 131255)

KEY TO FIGURE 1-3C

1. Anti-"G" Suit Tube Receptacle (Personal Gear Composite Disconnect to Suit)
2. Anti-"G" Suit Pressure Control
3. Microphone and Headset Plug (Personal Gear Composite Disconnect to Microphone and Headset)
4. Camera Compartment Temperature Indicator
5. Outside Air Temperature Indicator
6. Cameras Installed Indicator
7. Camera Compartment Heat Switch
8. Console Air Conditioning Outlet Control Handle
9. Ejection Seat Pre-ejection Lever
10. Wing Flaps Control Lever
11. Airplane Fuel System Control Panel
12. Engine Fuel System Control Panel
13. Strip Camera Control Panel
14. Canopy Emergency (Air) Control Lever
15. Canopy Normal (Hydraulic) Control Lever
16. Aft Bay Camera Control Panel
17. Forward Bay Camera Control Panel
18. K & CA Camera Power Switch
19. Oxygen Regulator
20. Catapult Grip
21. Throttle Friction Control Lever
22. Speed Brakes Circuit Breaker
23. Speed Brakes Control Switch
24. Stabilizer Position Indicator
25. Throttle and Microphone Button Switch
26. Stabilizer Power Control Panel
27. Location of Rudder Trim Tab Control Panel
28. Spare Lamps
29. Brake Emergency Control Handle
30. Speed Brakes Emergency Landing Override Switch
31. Oxygen Tube (Personal Gear Composite Disconnect to Mask)

NOTE

On F9F-6P airplanes with the interim flying tail modification installed, the stabilizer power control panel (item 26) is replaced by the stabilizer control selection panel shown as Detail BB in figure 1-2A. In addition, the stabilizer position indicator (item 24) is removed and replaced with an absolute stabilizer position indicator mounted on the instrument panel.

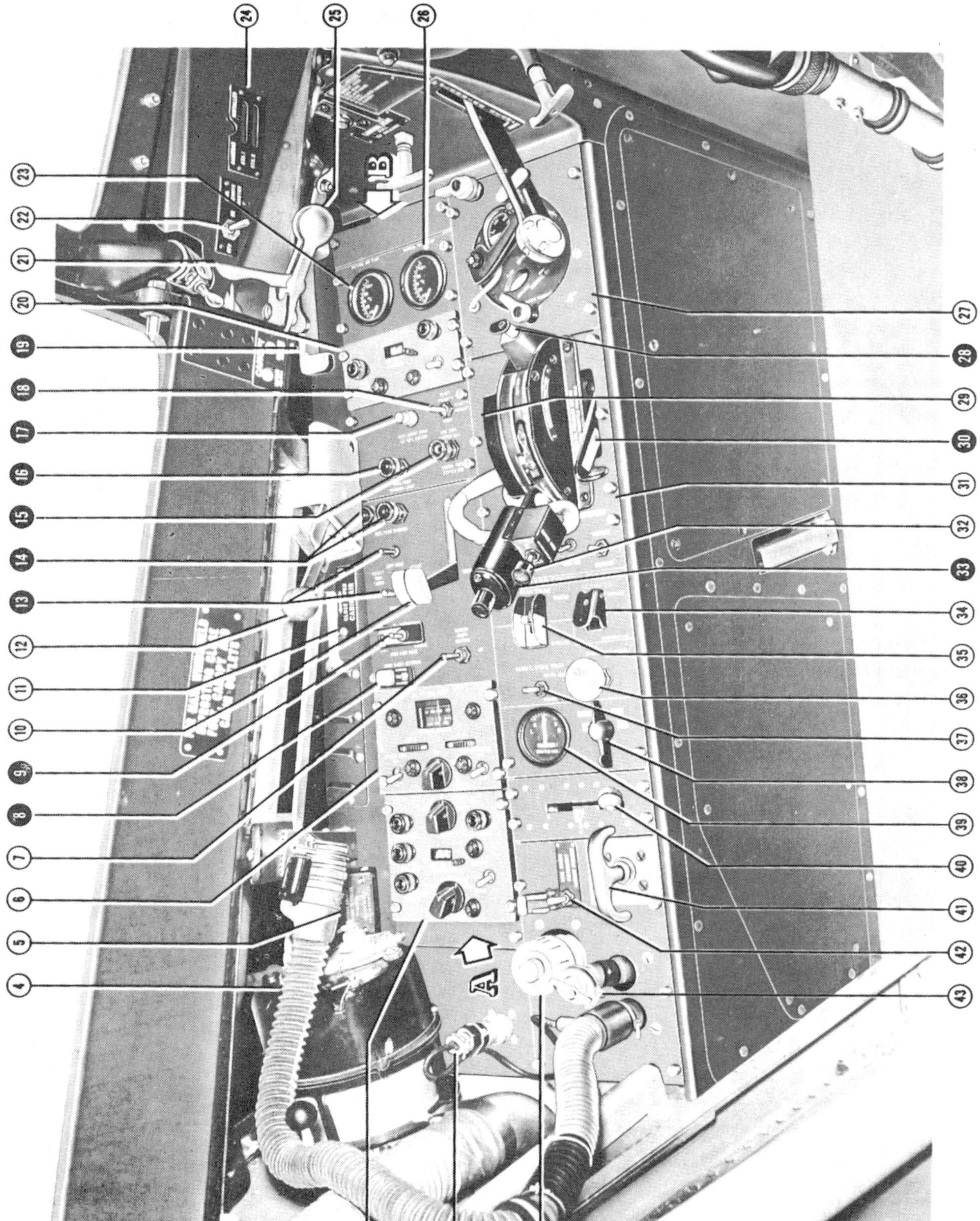

Figure 1-3C-1. Cabin Interior—Left Side (F9F-6P Airplanes ser No. 134446 and subsequent) (Sheet 1)

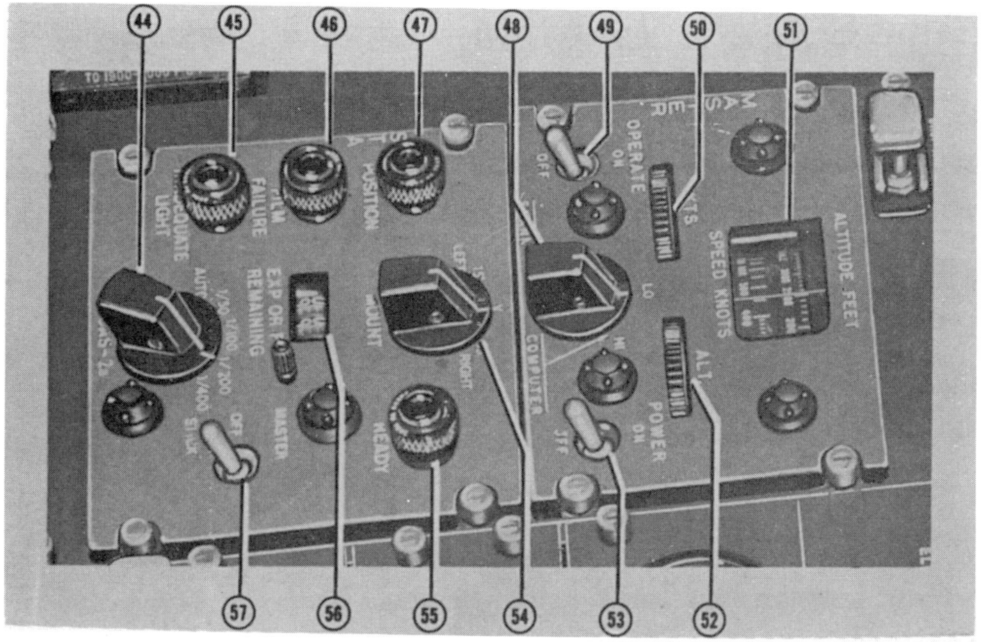

DETAIL A

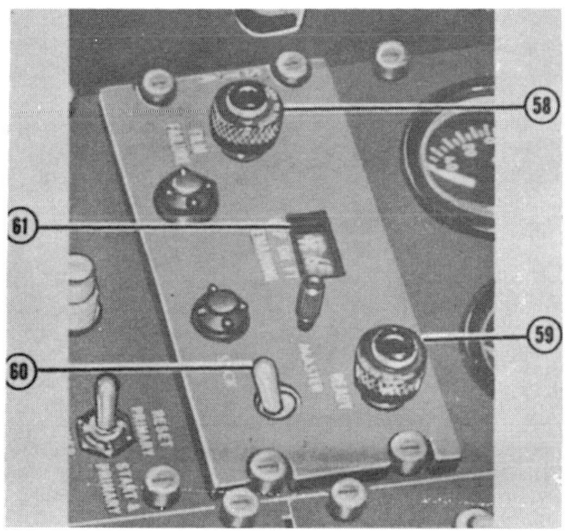

DETAIL B

1. "G" Suit Pressure Control
2. Microphone and Headset Plug
3. Station 2 Camera Control Panel
4. Oxygen Tube
5. Emergency Flaperette System Air Pressure Gage (behind oxygen tube)
6. Master Camera Control Panel
7. Camera Compartment Heat Control Switch
8. Drop Tanks Jettison Switch
9. Wing Tank Dump Switch
10. Wing Flaps Control Lever
11. Console Air Conditioning Outlet Control Lever
12. Pre-ejection Lever
13. Wing Tank Selector Switches
14. Wing Tank Fuel Flow Warning Lights
15. Emergency Fuel System Indicator Light
16. Engine Driven Fuel Pumps Warning
17. Fuel System Boost Pump Cut-off Switch
18. Engine Fuel System Selector Switch
19. Drop Tanks Jettison Emergency Control Lever
20. Station 1 Camera Control Panel
21. Canopy Emergency (Air) Control Lever
22. Camera Remote Master On-Off Switch
23. Outside Air Temperature Gage
24. Cameras Installed Card Case
25. Canopy Normal (Hydraulic) Control Lever
26. Camera Compartment Temperature Gage
27. Oxygen Regulator and Control Panel
28. Catapult Grip
29. Location of Speed Brakes Circuit
30. Throttle Friction Control Lever
31. Air Conditioning System Control Panel
32. Speed Brakes Switch
33. Throttle and Microphone Switch
34. Longitudinal Control System Selector Switch
35. Electric Trim Selector Switch
36. Emergency Manual Control Knob
37. Emergency Trim Switch
38. Rudder Trim Switch
39. Rudder Trim Position Indicator
40. Emergency Flaperette Power Control Lever
41. Brake Emergency Control Handle
42. Speed Brakes Emergency Landing Override Switch
43. "G" Suit Tube Receptacle
44. CAS-2a Camera Exposure Control Switch
45. Inadequate Light Indicating Light
46. Film Failure Indicator Light
47. Rotary Camera Mount Incorrect Position Indicator
48. Scanner - Computer Switch
49. Camera Operation Control Switch
50. Airspeed Setting Knob
51. Airspeed and Altitude Setting Indicator Dial
52. Altitude Setting Knob
53. Camera Power On-Off Switch
54. Camera Mount Position Selector Switch
55. Station 2 Camera Ready Indicator Light
56. Station 2 Camera Exposed Film or Film Remaining Counter
57. Station 2 Camera Mode Selector Switch
58. Station 1 Camera Failure Indicator Light
59. Station 1 Camera Ready Indicator Light
60. Station 1 Camera Mode Selector Switch
61. Station 1 Camera Exposed Film or Film Remaining Counter

Figure 1-3C-1. Cabin Interior—Left Side (F9F-6P Airplanes ser No. 134446 and subsequent) (Sheet 2)

Figure 1-3D. Cabin Interior—Forward (F9F-6P Airplanes ser No. 127473 through 131255)

KEY TO FIGURE 1-3D

1. Landing Gear Emergency (Air) Control Handle
2. Wheels and Flaps Position Indicator
3. Oil Pressure Indicator
4. Radio Altitude Low Limit Warning Light (See Note)
5. Elapsed Time Clock
6. K & CA Camera Alternate Power Switch
7. Low Fuel Boost Pressure Warning Light
8. Speed Brakes Position Indicator
9. Viewfinder Power Switch
10. Wide Angle-Drift Selector Switch
11. Filter Selector Switch
12. Radio Altimeter (See Note)
13. Grid Rotation Switch
14. Illumination Dual Control Rheostat
15. Camera Operating Indicator Lights
16. Viewfinder Eyepiece
17. Accelerometer
18. Strip Camera Synchronization Lights
19. Strip Camera Film Speed Indicator
20. Strip Camera Film Feed Light
21. Stand-by Compass
22. Fire Warning Lights
23. Fire Warning Light Circuit Test Switch
24. Low Fuel Level Warning Light
25. Fuel Quantity Indicator Press-to-Test Switch Button
26. Clock
27. Fuel Flowmeter
28. Arresting Hook Position Warning Light
29. Cabin Altimeter
30. Arresting Hook and Barrier Guard Control Handle
31. Arresting Hook Raising Switch Button
32. Air Conditioning Foot Outlet Control
33. Fuel Quantity Gage
34. Radio Compass Indicator
35. Rate of Climb Indicator
36. Turn and Bank Indicator
37. G-2 Remote Compass Indicator
38. Fast Erecting Gyro Horizon Indicator
39. Altimeter
40. Airspeed Indicator
41. Landing Gear Control, Down Lock Solenoid, Manual Release Knob
42. Airstart Emergency Igniter System Control Handle
43. Landing Gear Normal Control Handle and Landing Gear Unlocked Warning Light
44. Tailpipe Temperature Indicator
45. Tachometer

NOTE

Radio altimeter AN/APN-1 and low limit warning light are installed on airplanes ser No. 127473 through 127488. Radio altimeter AN/APN-22, which incorporates a low limit warning light, is installed on airplanes ser No. 127489 and subsequent.

On F9F-6P airplanes with the interim flying tail modification installed, an absolute stabilizer position indicator is installed on the upper left side of the instrument panel.

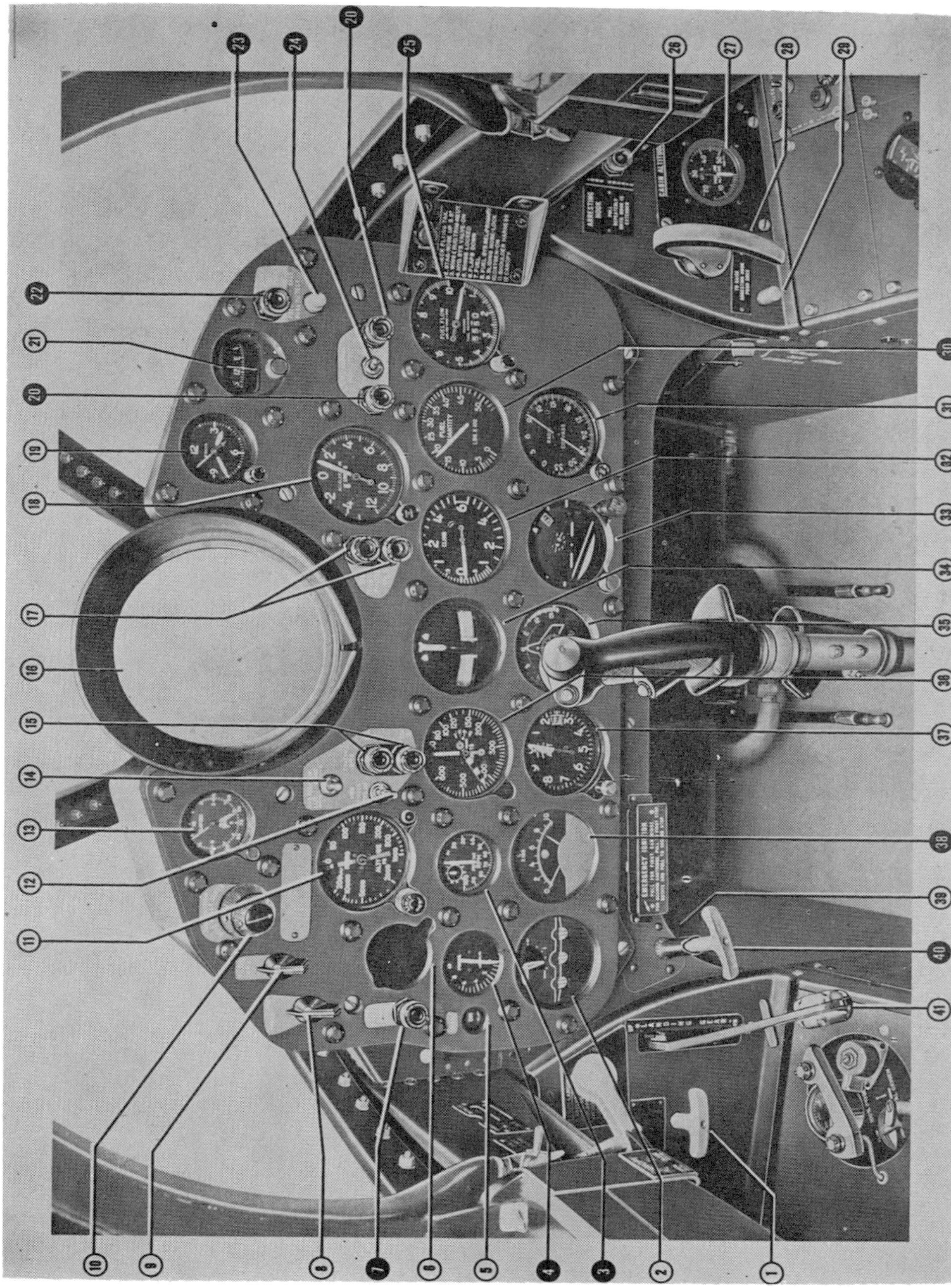

Figure 1-3D-1. Cabin Interior—Forward (F9F-6P Airplanes ser No. 134446 and subsequent)

KEY TO FIGURE 1-3D-1

1. Landing Gear Emergency (Air) Control Handle
2. Wheels and Flaps Position Indicator
3. Tachometer
4. Oil Pressure Indicator
5. Speed Brakes Position Indicator
6. Location of Absolute Stabilizer Position Indicator
7. Low Fuel Boost Pressure Warning Light
8. Wide Angle-Drift Selector Switch
9. Filter Selector Switch
10. Illumination Dual Control Rheostat
11. Radio Altimeter
12. Viewfinder Power Switch
13. Elapsed Time Clock
14. Grid Rotation Switch
15. Camera Operating Indicator Lights
16. Viewfinder Eyepiece
17. Film Failure Indicator Lights
18. Accelerometer
19. Clock
20. Fire Warning Lights
21. Stand-by Compass
22. Low Fuel Level Warning Light
23. Fuel Quantity Indicator Push-to-Test Switch Button
24. Fire Warning Light Circuit Test Switch
25. Fuel Flowmeter
26. Arresting Hook Position Warning Light
27. Cabin Altimeter
28. Arresting Hook and Barrier Guard Control Handle
29. Arresting Hook Raising Switch Button
30. Fuel Quantity Indicator
31. Radio Compass Indicator
32. Rate of Climb Indicator
33. Gyro Horizon Indicator
34. Turn and Bank Indicator
35. G-2 Remote Compass Indicator
36. Airspeed Indicator
37. Altimeter
38. Tailpipe Temperature Indicator
39. Location of Landing Gear Control, Down Lock Solenoid, Manual Release Knob
40. Airstart Emergency Igniter System Control Handle
41. Landing Gear Normal Control Handle and Landing Gear Unlocked Warning Light

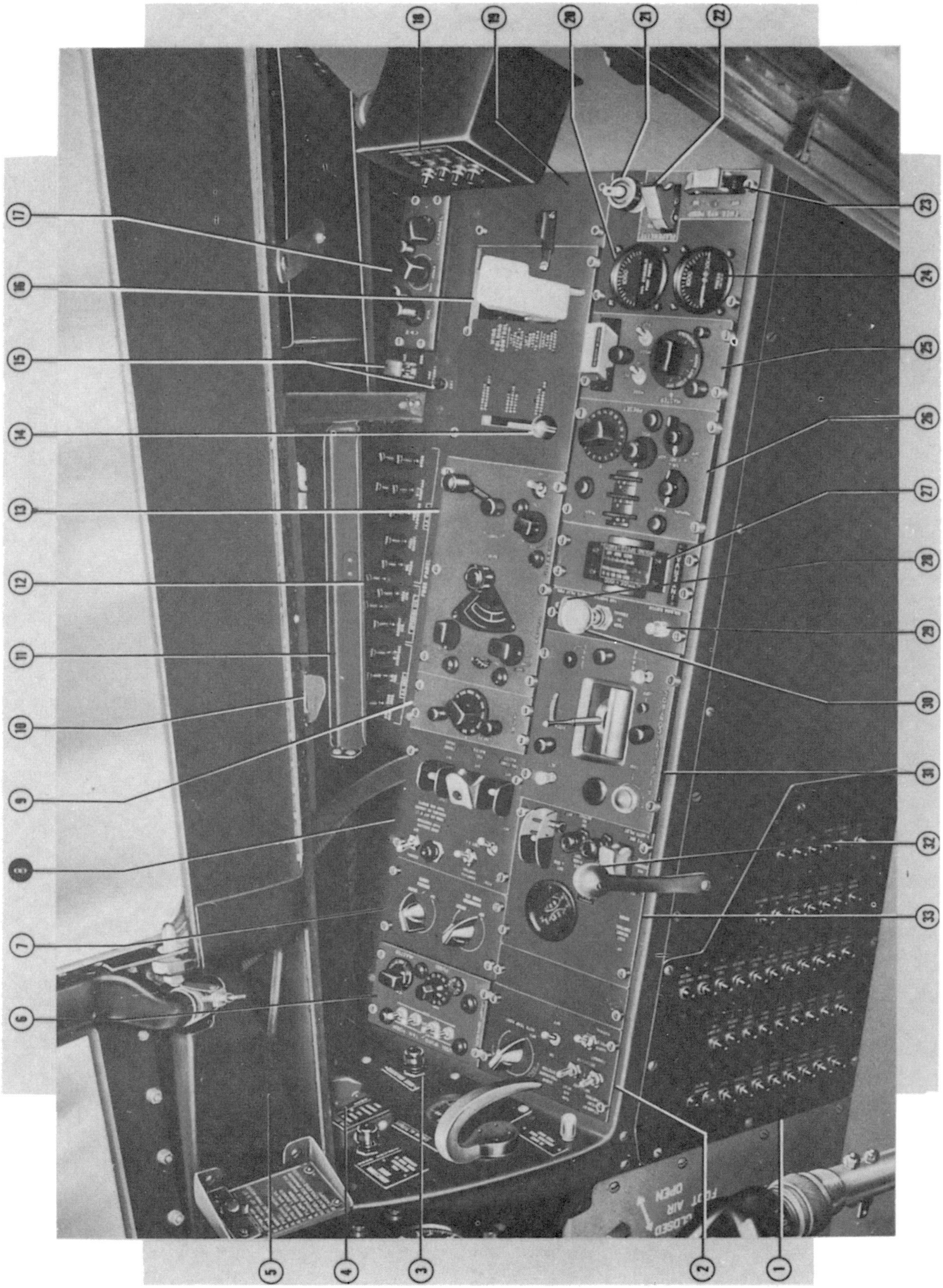

Figure 1-3E. Cabin Interior—Right Side (F9F-6P Airplanes ser No. 127473 through 131255)

KEY TO FIGURE 1-3E

1. Circuit Breaker Panel
2. Air Conditioning System Control Panel
3. Sliding Nose Unlocked Warning Light
4. Tail Skid Control Switch
5. Location of Viewfinder Circular Slide Rule
6. Exterior Lights Control Panel
7. Interior Lights Control Panel
8. Engine and Miscellaneous Control Switches
9. Radio Altimeter Limit Switch (See Note)
10. Chartboard Stowage Latch
11. Chartboard Stowage Channel
12. Fuse Panel
13. AN/ARN-6 Radio Compass Control Panel
14. Hydraulic Pressure ON-OFF Control Lever
15. Antenna Relay Control Switch and Circuit Breaker
16. Control Modified, See Figure 1-10
17. AN/ARR-2A Navigational Radio Control Panel
18. Stabilizer Trim Power Circuit Breaker Panel
19. Spare Lamps and Fuses
20. Auxiliary Hydraulic Pump Pressure Gage
21. Utility Receptacle
22. Flaperette System Hydraulic Pump Control Switch
23. Auxiliary Hydraulic Pump Switch
24. Main Hydraulic System Pressure Gage
25. AN/APX-6 IFF Control Panel
26. AN/ARC-27 UHF Command Set Control Panel
27. Interval Control Computer
28. Auto Pilot Power Switch
29. Holding Switch
30. Yaw Damper Switch
31. Auto Pilot Control Panel
32. Seat Height Control Lever
33. Electrical System Control Panel

NOTE

Radio altimeter AN/APN-1 limit switch is installed on airplanes ser No. 127473 through 127488.

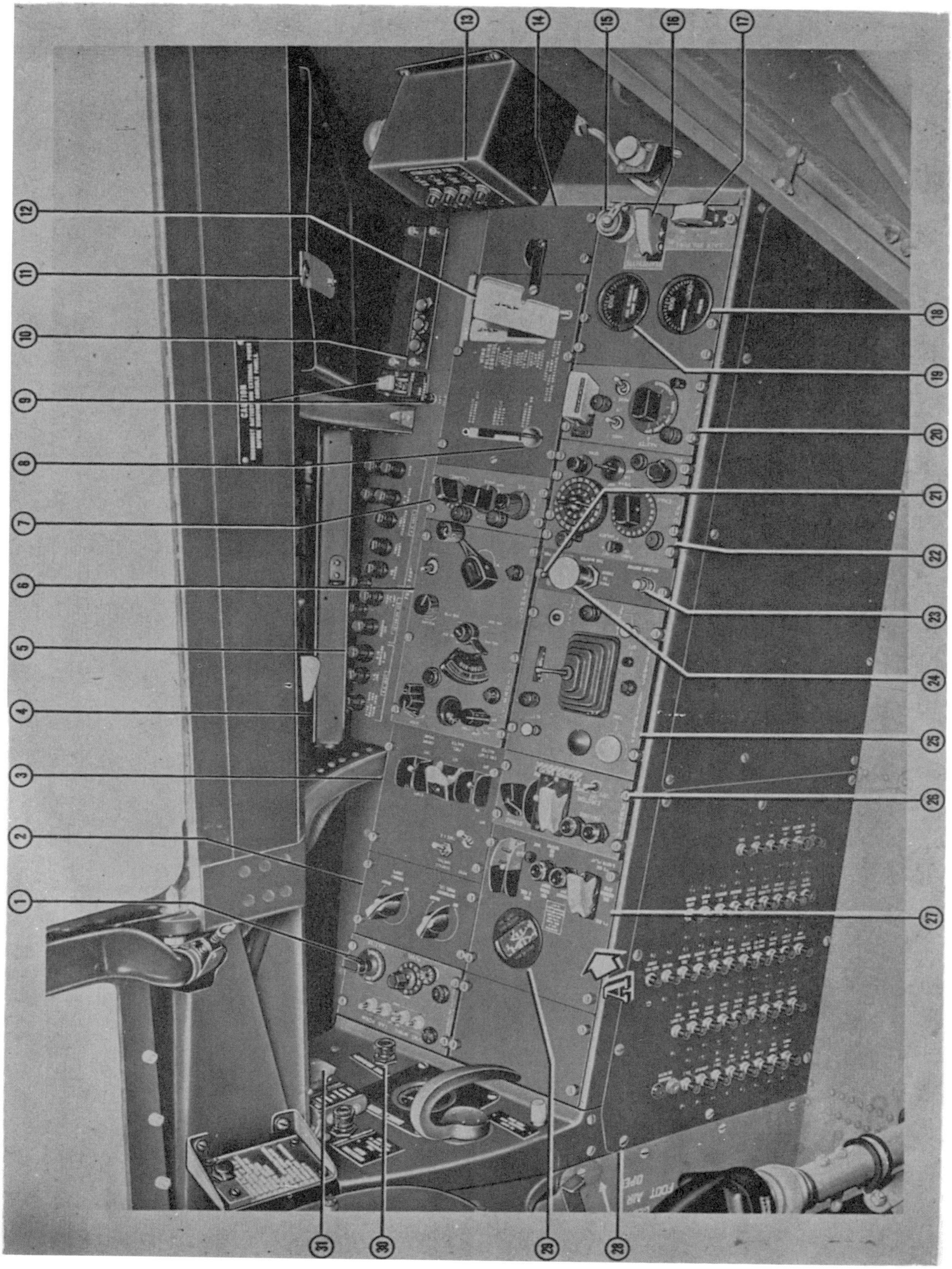

Figure 1-3E-1. Cabin Interior—Right Side (F9F-6P Airplanes ser No. 134446 and subsequent) (Sheet 1)

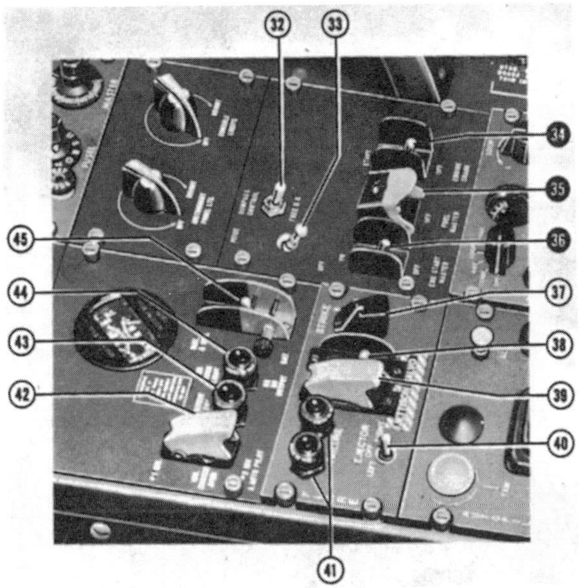

DETAIL A

1. Exterior Lights Control Panel
2. Interior Lights Control Panel
3. Engine and Miscellaneous Control Switches
4. Chartboard Stowage Compartment
5. Fuse Panel
6. Radio Compass Control Panel
7. Navigational Radio Control Panel
8. Hydraulic Pressure Control Lever
9. Antenna Relay Control Switch and Circuit Breaker
10. Spare Lamps Panel
11. Map Case
12. Control modified, see figure 1-10.
13. Stabilizer Trim Power Circuit Breaker Panel
14. Spare Lamps and Fuse Stowage Compartment
15. Utility Receptacle
16. Flaperette System Control Switch
17. Auxiliary Hydraulic Pump Control Switch
18. Main Hydraulic System Pressure Gage
19. Auxiliary Hydraulic Pump Pressure Gage
20. IFF Control Panel
21. Auto Pilot Power Control Switch
22. UHF Command Set Control Panel
23. Auto Pilot Holding Switch
24. Yaw Damper Control Knob
25. Auto Pilot Control Panel
26. Flare Release Control Panel
27. Electrical System Control Panel
28. Main Circuit Breaker Panel
29. Voltammeter
30. Nose Unlock Warning Light
31. Tail Skid Control Switch
32. G-2 Compass Control Switch
33. Pitot Heat Control Switch
34. Engine Cranking Switch
35. Fuel Master Switch
36. Engine Start Master Switch
37. Strike Camera Control Switch (Not Used)
38. Flare Emergency Salvo Switch
39. Flare Pad Emergency Jettison Switch
40. Flare Ejector Switch
41. Flare Release Failure Indicator Lights
42. Inverter Changeover Switch
43. Inverter Warning Light
44. Generator Warning Light
45. Battery Switch

Figure 1-3E-1. Cabin Interior—Right Side (F9F-6P Airplanes ser No. 134446 and subsequent) (Sheet 2)

Note

1. Seat height is controlled electrically on F9F-6 airplanes ser No. 128240 and subsequent and F9F-6P airplanes ser No. 128307 and subsequent.
2. Emergency Ejection Seat Arming Control Handle installed on F9F-6 airplanes ser No. 127400 and subsequent.

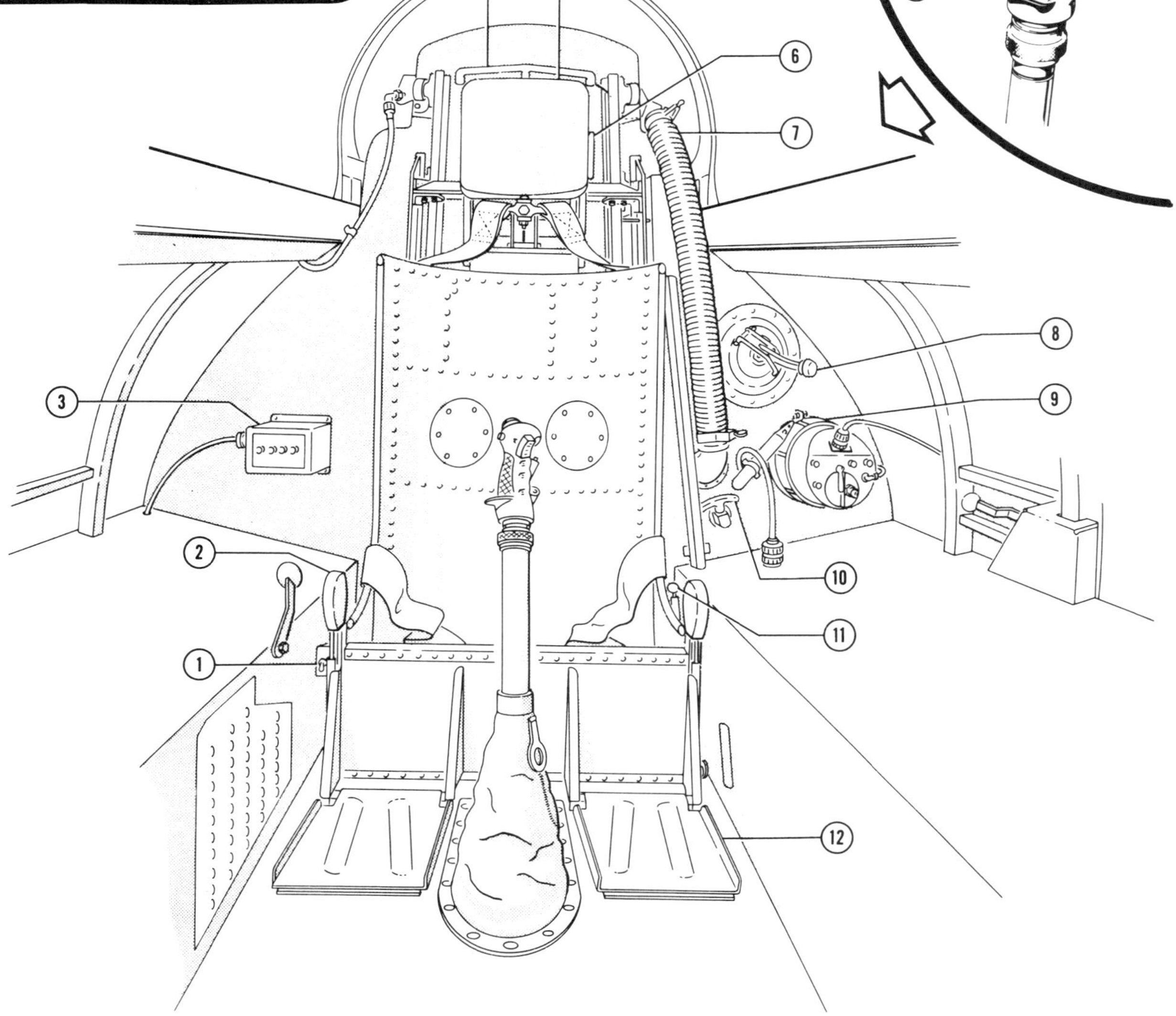

1. Electric Seat Height Control (See Note 1)
2. Ejection Seat Knee Brace (2)
3. Stabilizer Power Circuit Breaker Panel
4. Ejection Seat Face Cover Handle
5. Headrest
6. Emergency Ejection Seat Arming Control Handle (See Note 2)
7. Canopy Defrosting Tube
8. Cabin Pressure Dump Valve Control Lever
9. Cabin Pressure Regulator
10. Speed Brake Manual Override Control Handle
11. Shoulder Harness Inertia Reel Lock Control Lever
12. Ejection Seat Foot Rest (2)
13. Auto Pilot Emergency Off Switch (F9F-6P Only)
14. Gun Trigger Switch (F9F-6 Airplanes) Camera Trigger Switch (F9F-6P Airplanes)
15. Bomb Release Switch (F9F-6 Airplanes) Extra Picture Switch (F9F-6P Airplanes)
16. Horizontal Stabilizer Trim and Wing Trim Control Button
17. Rocket Switch (F9F-6 Airplanes) P-2 Strike Camera Switch (F9F-6P Airplanes)

Figure 1-4. Cabin Interior—Aft

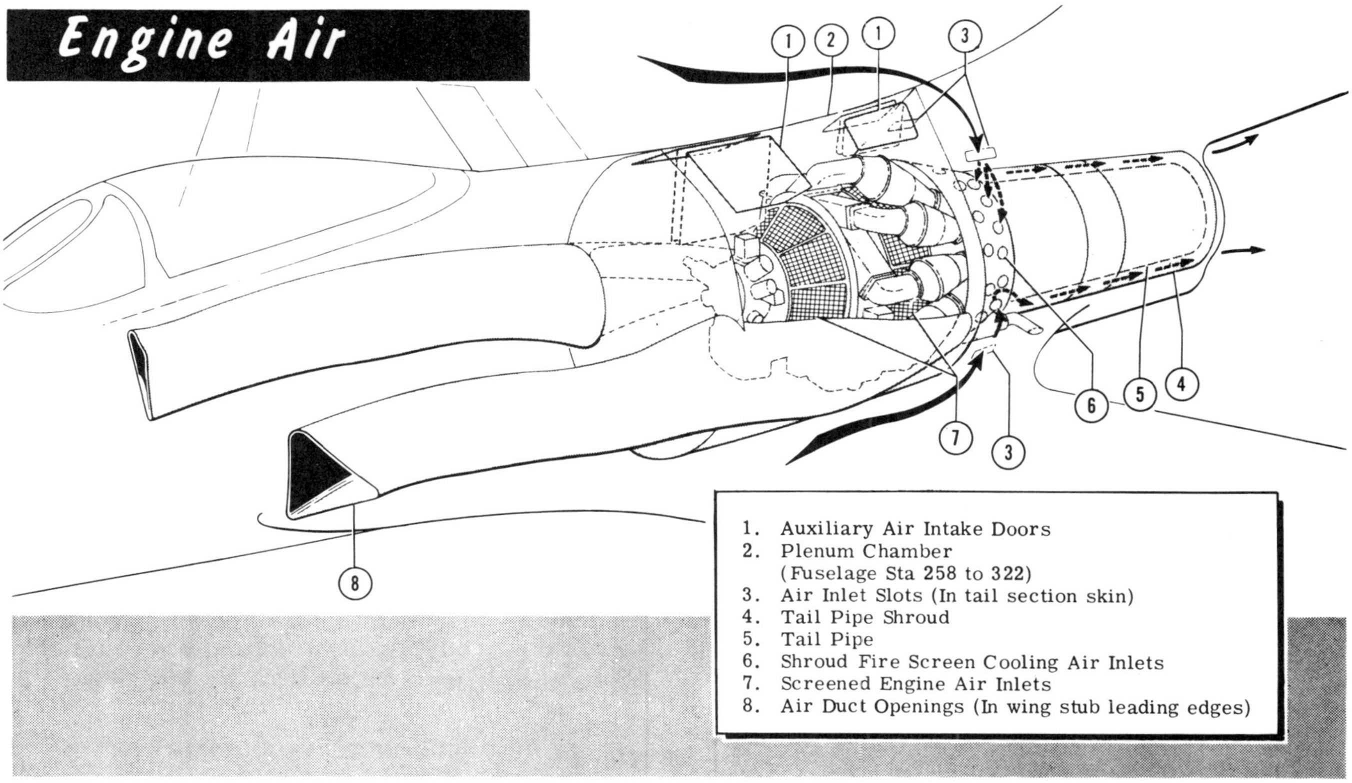

Figure 1-5. Engine Air System

air through the rear screen support and forces it rearward to cool the center and rear bearings. Part of this air passes out radially to cool the turbine disc front face, thence into the cooling air manifold and out the cooling air duct at the fuselage bottom. The second system provides cooling air for the tailpipe. Air is drawn in through three flush type ram air scoops in the tail section skin, circulates between the tailpipe and the shroud and flows out at the rear.

IGNITION SYSTEM.

The ignition system is used only during engine starting. When the turbine has reached sufficient speed to permit engine light-up, the two igniter plugs (in combustion chambers No. 4 and 7) are energized by the throttle quadrant limit switch. The engine starting master switch must be on in order for the ignition circuit to be operative for both ground and airstarts. The ignition system will be energized for 60 seconds after the limit switch is actuated, after which it will shut off automatically. However, when the engine start master switch is turned off to interrupt the starting cycle, the ignition system will also be shut off.

IGNITION CONTROLS.

Movement of the throttle (22, figure 1-2A) outboard into the START detent actuates the ignition system limit switch, thus allowing current to be supplied to the ignition units. Ignition cut-off is accomplished automatically in 60 seconds by a time delay switch. The ignition circuit will not operate unless the engine start master toggle switch, located on the right console, is set forward to ON. While it is normally on at all times during flight, it should be noted that ignition for airstarts cannot be obtained with the engine start master switch at OFF.

ENGINE STARTING SYSTEM.

The starting system is designed to rotate the engine at the speed required to obtain a successful engine light-up when fuel and ignition are introduced in the combustion chambers.

Note

An external electrical power supply must be used for ground starts.

The starter is not used for airstarts, as airspeed may be controlled by the pilot to maintain the proper engine windmilling speed for an engine light-up.

ENGINE STARTING CONTROLS.

ENGINE START MASTER SWITCH.

This two position, guarded toggle switch marked ON and OFF is located on the right console (45, figure 1-3B). It operates in conjunction with the engine cranking switch and the limit switch at the START position on the throttle to control the starter and ignition circuits, respectively. The engine start master switch must be set to ON before the starter and ignition circuits can be energized by their respective control switches. Setting the switch to OFF removes power from the starter and ignition circuits when completion of a starting cycle is not desired (false start).

Note

For airstarts, the engine start master switch must be set to ON to enable the ignition system to be energized when the throttle is moved to START.

ENGINE CRANKING SWITCH.

This momentary switch, equipped with a guard, is on the right console electrical control panel, outboard (43, figure 1-3B). With the engine start master switch ON, the engine cranking switch is set forward to START to energize the starter for ground starts or for clear engine when the engine has failed to start and unburned fuel must be cleared. In this latter operation, the throttle is set at CLOSED. The engine cranking switch circuit is broken when the engine start master switch is OFF.

AIRSTART EMERGENCY IGNITER SYSTEM CONTROL.

The engine is equipped with an airstart emergency igniter system, which consists of a "T" handle control in the cabin connected by cables to the firing pins for two cartridges installed in combustion chambers No. 4 and 7. The "T" handle control is mounted on the bottom left side of the instrument panel (43, figure 1-3A). Pulling the handle aft to its first stop fires a single cartridge to provide emergency ignition for an airstart attempt. After firing the first cartridge, the second may be fired by pulling the handle aft to its first stop, rotating it 90 degrees in either direction and then pulling full aft.

Note

This control is spring loaded to return to its normal position after firing the first cartridge. To fire the second cartridge for a second airstart attempt the handle must be pulled to the first stop, rotated 90 degrees in either direction, and then pulled out to its second stop. If the control is pulled through as described above, in a continuous motion, both cartridges will be fired in quick succession.

The emergency igniter system is provided as an additional safety device for use in case of failure to relight with the normal airstarting ignition system. It is important to note that only two ignition attempts are available with this emergency igniter system. The airstart emergency igniter system control is rigged so that the cartridges may be fired singly or in rapid succession, if desired. Therefore, these emergency igniters should be used only when attempts to ignite with the normal airstarting igniter system have failed or when time permits only one, or at the most two possible relighting attempts. Conditions should be set as nearly as possible to those outlined in Section III before attempting starts with emergency igniters. All starting attempts should be made with a single igniter except under extreme circumstances when time will permit only one attempt. In this case both igniters should be fired in rapid succession.

ENGINE INDICATORS.

TACHOMETER INDICATOR.

The tachometer is on the main instrument panel to the left of the center line. It indicates engine rpm in percentage of military rpm (100%=11000). The dial is numbered in tens from 0 to 100. A small pointer rotating on a sub-dial shows graduations of one per cent of engine rpm to provide more accurate reading than the large pointer (45, figure 1-3A).

TAILPIPE TEMPERATURE INDICATOR.

This indicator, on the main instrument panel to the left of the center line, shows tailpipe temperature in degrees centrigrade times 100. It is numbered from 0 to 10 in increments of two (45, figure 1-3A). There are two pointers on the dial face of the instrument. The left one is the indicating pointer and the right one the reference pointer. The latter pointer may be set to any position by turning the screw at the bottom of the dial face. The clockwise movement of the indicating pointer is stopped by the reference pointer. When the indicating pointer contacts the reference pointer, a warning light incorporated in the indicator dial face will glow.

Note

The reference pointer should always be set at the highest tailpipe temperature limit (800°C), as the indicating pointer cannot indicate readings above the setting of the reference pointer.

FIRE WARNING INDICATOR LIGHTS.

Two red press-to-test fire warning lights, with a warning light circuit test toggle switch, are located on the top of the instrument panel to the right of the center line. The lights are energized by fire warning detectors located at various points around the engine to warn of engine compartment fire. The lights will glow when the fire warning test switch is actuated, indicating that the fire warning light circuit is in working order (16, 18, figure 1-3A).

OIL SYSTEM.

The engine oil system is a self-contained high pressure system which provides pressure oil feed to the three main bearings and the accessory drives. The bulk of the oil is contained in an oil tank located on the inner side of the front air inlet screen support. The oil sump contains a three-section, engine driven oil pump, four oil strainers and provisions for the oil pressure gage connections. The middle section of the pump draws oil from the tank and forces it through tubing to nozzles which lubricate the accessory case drives and the front, center, and rear bearings. Engine drain oil from the center and rear bearings is scavenged by the upper section; and from the front bearing, accessory drives and pressure oil relief valve by the lower section. The oil from both pump sections is combined and returned to the tank through a single line. The oil reservoir is filled through a fillerneck on the fuselage right side, accessible through a filler cap below the auxiliary air

FUEL QUANTITY DATA

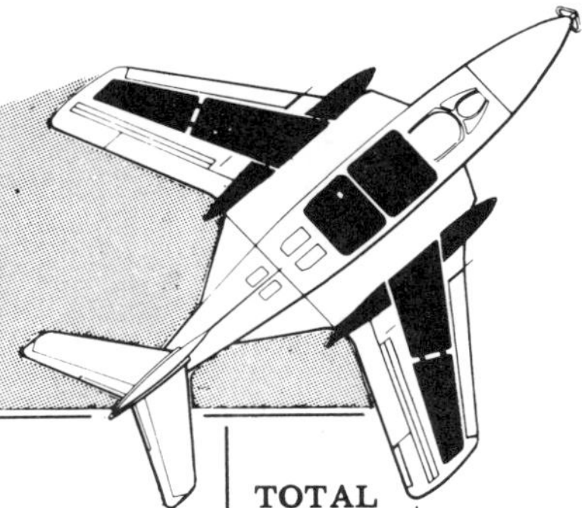

Note

(JP-3 or JP-4) 6.5 Lb/Gal
(AVGAS) 6.0 Lb/Gal

TANKS	NO.	USABLE FUEL LEVEL FLIGHT (each)			UNUSABLE FUEL LEVEL FLIGHT (each)			EXPANSION SPACE (each)	TOTAL VOLUME (each)
		US Gallons	Lb JP-3 or 4	Lb AVGAS	US Gallons	Lb JP-3 or 4	Lb AVGAS	US Gallons	US Gallons
Forward Fuselage	1	506	3289	3036	24	156	144	16	546
Aft Fuselage	1	240	1560	1440	0	0	0	7	247
Wing	2	78	507	468	0	0	0	5	83
Droppable Tanks	2	150	975	900	0	0	0	6	156

Figure 1-6. Fuel Quantity Data Table

door. The oil system capacity is 26 US pints with a foaming space of approximately 20 US pints. See figure 1-2 for oil grade and specification.

WARNING

Inverted flight is limited to 10 seconds by the engine oil system.

OIL PRESSURE INDICATOR.

The oil pressure indicator is on the left side of the main instrument panel (6, figure 1-3A). Its range is from 0 to 100 psi. The desired oil pressure operating range is from 10 psi minimum to 45 psi maximum.

AIRPLANE FUEL SYSTEM.

The forward (main) tank and the rear tank are of the soft, self-sealing type and are located between the cabin and the engine. Two interconnected non-sealing bladder type tanks run the length of the wing outer panel and are located between the front and main beams. An electrical system is provided for dumping the fuel from the wing tanks in an emergency. See figure 1-2 for fuel grade and specification. The fuselage tanks are pressurized by ram air let into the tanks from a scoop on the fuselage bottom. The wing tanks are pressurized from ram air inlets at the wing tip leading edges which are equipped with spring loaded valves. During fuel dumping, these inlet valves are opened to permit full ram air pressure and the dump valves are opened to permit jettisoning through the dump nozzles.

On airplanes[1] equipped for external stores, a 150 gallon drop tank may be supported by each bomb rack. Each drop tank is pressurized through a curved, scarfed vent tube protruding up through the top of the tank just forward of the support lugs.

This is a constant feed system, i.e., fuel flows from the wing and/or drop tanks and the rear fuselage tank to the forward fuselage tank. A check valve in the line connecting the fuselage tanks prevents reverse flow from the forward to the rear tank. A check valve in each drop tank line prevents fuel flow from the respective wing tank in to the drop tank. The fuel level in the fuselage tanks is approximately the same until the rear tank is emptied,

[1] F9F-6 airplanes ser No. 128145 and subsequent and F9F-6P airplanes ser No. 130768 and subsequent.

MAXIMUM ATTAINABLE % RPM vs FUEL QUANTITY

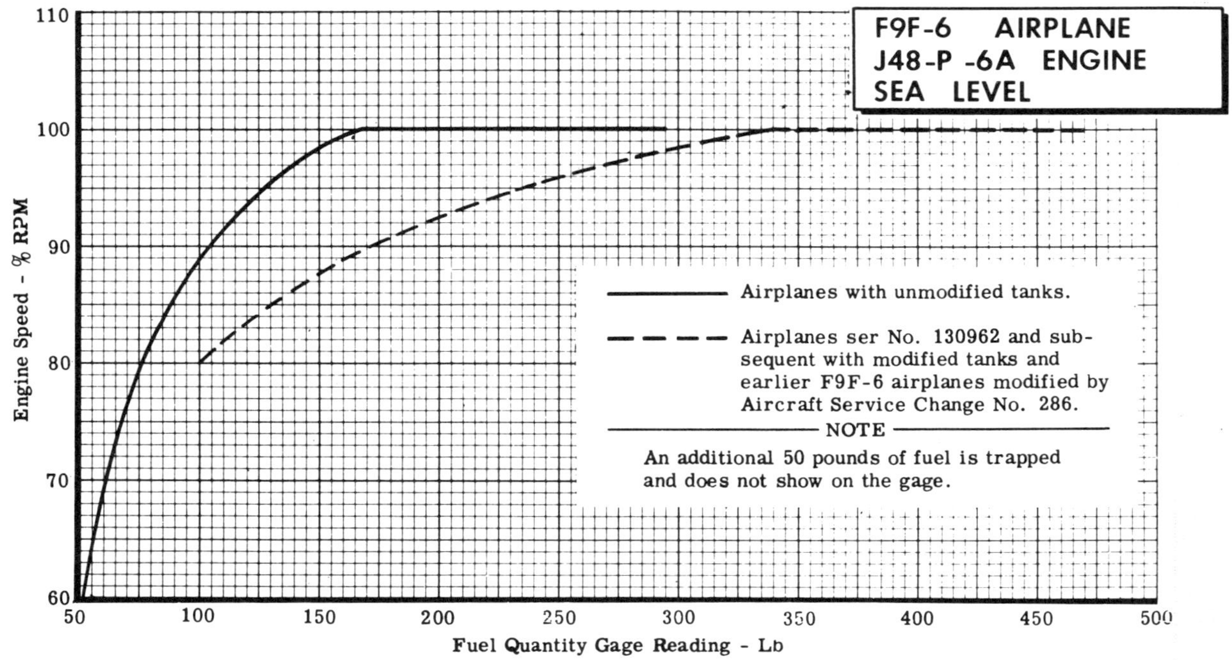

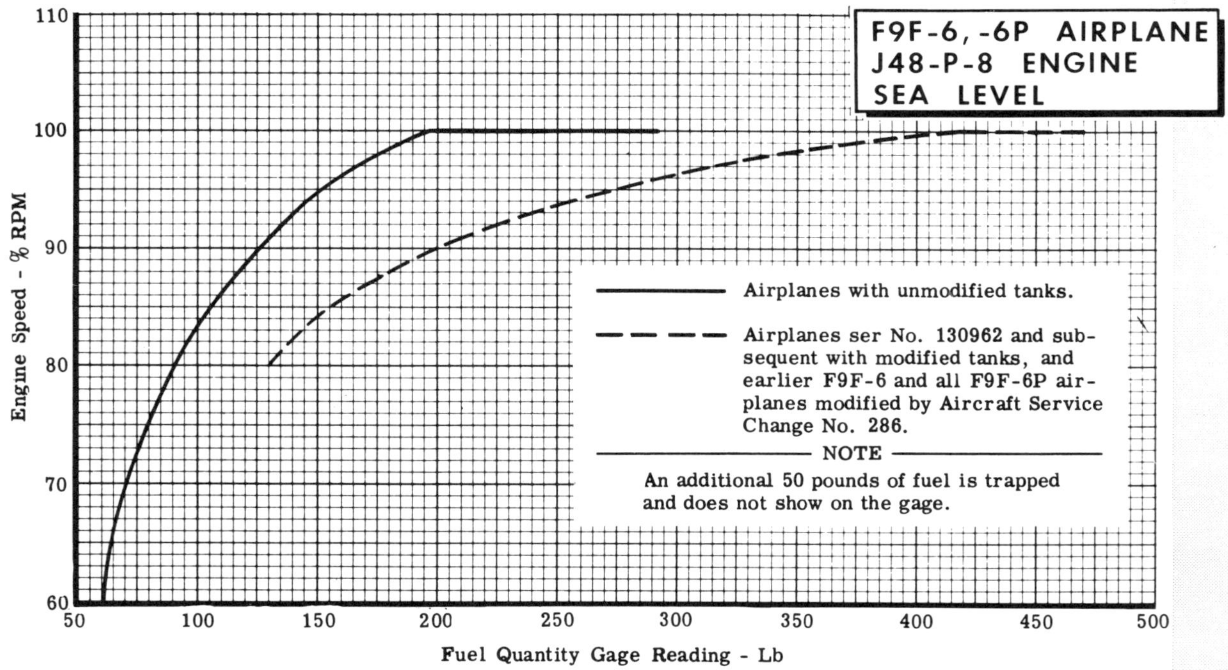

Figure 1-6A. Maximum Attainable Percent RPM vs Fuel Quantity Curves

at which time approximately 100 gallons (650 lb JP-3, -4; 600 lb AVGAS) remain in the forward tank. All fuel feed to the engine is from the forward tank.

There are certain limitations to flying with low fuel quantities. Figure 1-6A shows the maximum attainable engine speed vs indicated fuel quantity. The minimum fuel quantity gage reading (figure 1-6A) necessary for 100% engine rpm on airplanes with tanks not modified by F9F Aircraft Service Change No. 286 is 340 pounds for J48-P-6A engines and 410 pounds for J48-P-8 engines. The minimum fuel quantity gage reading necessary for 100% engine rpm on later airplanes[1] and on airplanes with tanks modified by F9F Aircraft Service Change No. 286 is 165 pounds for J48-P-6A engines and 195 pounds for J48-P-8 engines.

The unusable fuel quantities listed in figure 1-6 are based on 50 pounds of trapped fuel plus the minimum amount of fuel required to obtain 83% engine rpm, which is the engine speed (approximate) required for carrier approach. See the Low Fuel Quantity paragraph in Section III.

Note

> The engine fuel pumps warning light will be on while flying with low fuel quantities.
>
> Because the loss of thrust is gradual as fuel is used, some warning is provided.

The fuel quantities discussed above and shown in figure 1-6A represent fuel quantity gage readings in level flight, including level coordinated turns. Higher rpm for the same fuel quantity will be obtained for the J48-P-6A engine or for the J48-P-8 engine on a warm day or at higher altitudes because of the lower fuel flow required corresponding to the lower thrust.

NORMAL OPERATION.

When the fuel system master switch is ON, the shut-off valve is opened and the tank boost pump sends fuel to the engine at 6 to 25 psi. When a wing or drop tank is selected, a transfer pump sends fuel from that tank into the forward fuselage tank, provided the circuit to the transfer pump has been armed by action of the float switch in the forward tank when space is available in that tank. A red warning light glows when either wing tank has been emptied, provided the fuel level in the forward tank is below full and the transfer pump is operating. On airplanes equipped for external stores, a fuel flow indicator switch, located upstream of the transfer pump in each wing fuel line, will illuminate the warning light when fuel flow through that line drops to less than 1.0 gpm. If the forward tank is full, its float switch will not permit the transfer pump to operate and the lights will not glow, having been deenergized temporarily by pump stoppage. A seven gallon reservoir in the bottom of the forward tank maintains constant feed for inverted flight or maneuvers for approximately 10 seconds when 90 gallons (585 lb JP-3, -4; 540 lb AVGAS) of fuel remain in the tank. With less than 90 gallons, there will be proportionately less fuel in the reservoir. Fuel quantity in the fuselage tanks is measured by capacitance units. A fuel flowmeter shows rate of flow and amount of fuel remaining (both in pounds) in the system.

FUEL SYSTEM CONTROLS.

WING TANK SELECTOR SWITCHES.

Two switches on the fuel control panel of the left console (30, figure 1-2A) control the transfer pump. The outboard switch is for the left wing tank, the inboard for the right tank. Moving either switch forward to the TRANS position turns on the electric transfer pump which sends fuel from the selected wing tank to the forward fuselage tank. If the forward tank is full, a float switch will prevent the transfer pump from operating. Setting either switch aft to CLOSE stops transfer of fuel from the selected tank by shutting off the wing tank shut-off valve. When a tank shows empty, its transfer switch should be set to CLOSE. Both switches must be set to CLOSE to turn off the electric transfer pump.

On airplanes equipped for external stores, the selector switches are identified left wing and right wing, and each has three positions; WING TANK, OFF, and DROP TANK. Moving either switch out of the OFF position sets a wing tank selector valve to the selected wing or drop tank position, opening the line for fuel flow, and turns on the transfer pump. Final stages of transfer will result in a passage of air and fuel which will cause intermittent operation of the fuel flow switch and a corresponding flicker in the warning light. This is a normal condition. When either light glows steadily, the associated switch should be set to OFF. The other switch should be left in the selected position until the light glows steadily again. Failure to turn either switch to OFF when its light glows may result in leaving some fuel in one tank, as the transfer pump will suck air from the empty tank. This will decrease the flow from the tank containing fuel, possibly below the flow indicator switch setting of 1.0 gpm, and result in a false indication of empty.

WING TANK DUMP SWITCH.

This two position switch is mounted in a channel type guard on the fuel control panel of the left console (28, figure 1-2A). Movement of the switch outboard to DUMP actuates an electric motor which opens the dump valves in both wing tanks and allows ram air to force the fuel out the nozzles into the airstream. This switch is left at the NORMAL position until it is desired to dump wing tank fuel or to purge the tanks after wing tank fuel has been exhausted. In either case, after the switch has been set to DUMP, it must be left there to ensure complete purging, but must be reset to NORMAL before landing.

The airplane must be in a nose-up attitude to dump wing tank fuel. The fuel dump outlet is located at the

[1] F9F-6 airplanes ser No. 130962 and subsequent.

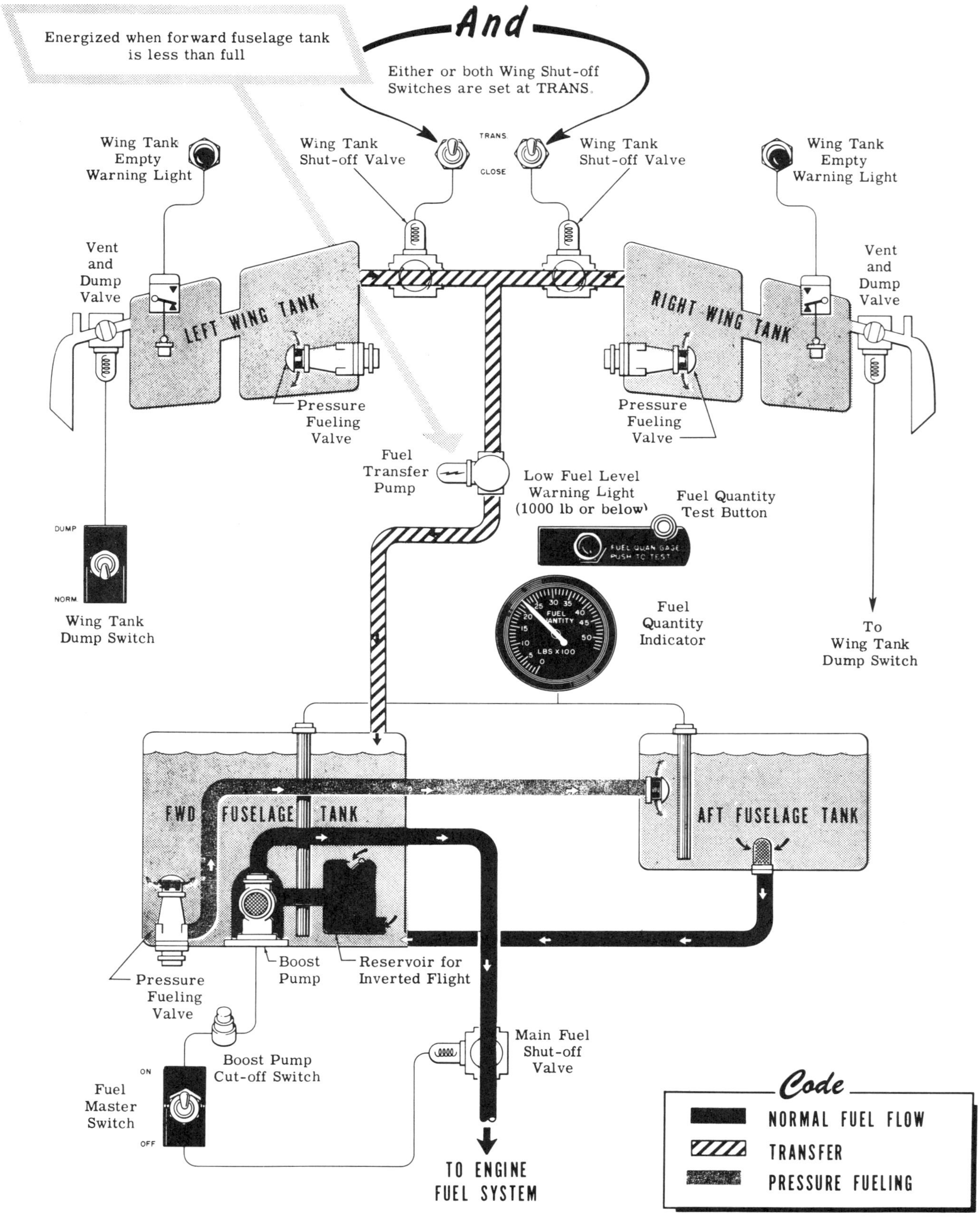

Figure 1-7. Airplane Fuel System Schematic Diagram (Sheet 1)

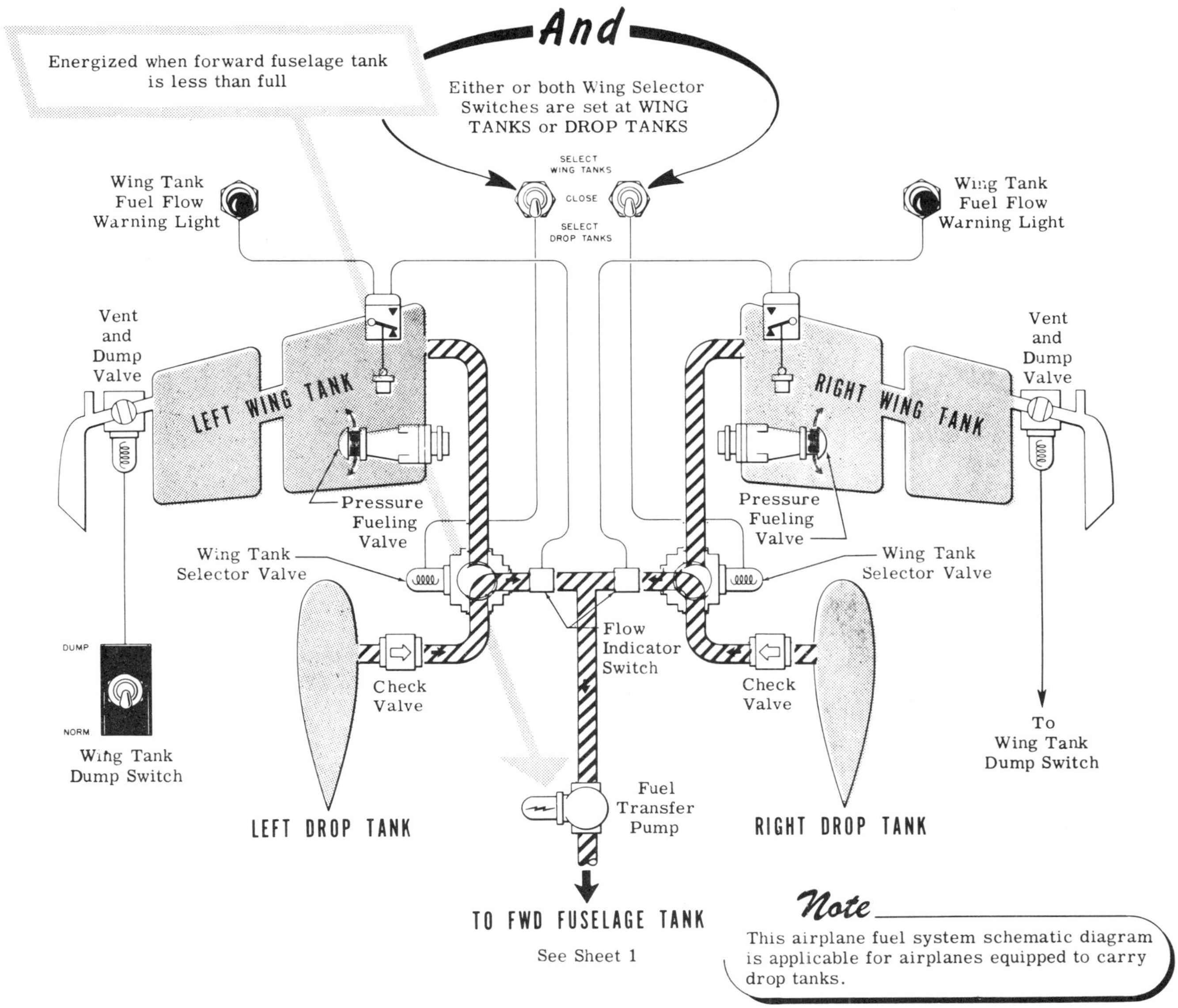

Figure 1-7. Airplane Fuel System Schematic Diagram (Sheet 2 of 2)

extreme aft outboard corner of the main wing cell. To insure complete fuel flow to the dump valves, it is necessary to fly the airplane in a nose-up attitude since the wing has a zero degree angle of incidence. This places the aft outboard corner of the cells lower than the leading edge and inboard portions.

Note

When airplane is decelerating, inertia forces will prevent fuel from being dumped from wing tanks.

FUEL SYSTEM INDICATORS.

Also see Engine Fuel Control System.

FUEL QUANTITY INDICATOR.

The quantity indicator, on the right side of the main instrument panel (28, figure 1-3A), shows quantity of fuel in pounds. The dial is marked 0—53 and indicates pounds times 100. The indicator is operated by electric capacitance units in the front and rear tanks (there is no quantity indication for the wing tanks). A push button (17, figure 1-3A) is installed above the indicator to check operation of the system. With the battery switch set to BAT. & GEN., pushing this button will cause the indicator pointer to move to the low end of the dial. If the system is operating properly, the pointer will return to its original indication when the button is released.

Note

Fuel varies in weight per gallon, dependent upon its specific gravity and temperature. Therefore, the notation FULL does not appear on the indicator dial and the pilot should anticipate variations in the instrument readings when the fuel tanks are full.

LOW FUEL LEVEL WARNING LIGHT.

This red light, on the right side of the main instrument panel (19, figure 1-3A), glows when approximately 1000 pounds of fuel remain in the system.

WARNING

Descents with the fuel low level warning light on must be made in accordance with the procedure given in Appendix I, Description of Charts and Tables paragraph, step h.

WING TANKS EMPTY WARNING LIGHTS.

Warning lights for each wing tank (29, figure 1-2A), operated by float switches, glow when the tanks are empty. On airplanes equipped for external stores, the lights are operated by fuel flow switches located in the wing fuel lines. The lights will flicker intermittently in the final stages of fuel transfer from either wing, and glow steadily when fuel flow through the line drops to less than 1.0 gpm.

FLOWMETER.

The fuel flowmeter is on the right side of the main instrument panel (21, figure 1-3A). The indicator has a pointer which shows the rate of flow (in pounds per hour) and a counter which shows the amount of fuel remaining (in pounds). The counter must be reset at each refueling to indicate total fuel aboard at the start of a flight.

ELECTRICAL SYSTEM—F9F-6 AIRPLANES.

The basic system is a 28 volt, direct current, open wire system which uses the aircraft structure for a ground. Power is provided by a combination a-c—d-c generator and a battery. These, plus an a-c voltage regulator, a d-c voltage regulator, a reverse-current relay, wiring and bus bars, make up the main supply circuit. The controls, with the exception of armament, engine and certain special items, are on the right console. See figure 1-8 for additional information on the electrical system.

VARIABLE FREQUENCY A-C POWER.

The combination a-c—d-c generator supplies 115 volt, variable frequency (800-1600 cps), single phase a-c power directly to the AN/APX-6 IFF and AN/APG-30 gun ranging radar equipment. An a-c voltage regulator is provided to maintain the generator output at 115 ±3 volts. The AN/APX-6 and AN/APG-30 variable frequency a-c circuits are each protected by a circuit breaker mounted on the main circuit breaker panel.

Note

When the generator warning light glows (see Generator Warning Light paragraph), it indicates the loss of variable frequency a-c as well as d-c power.

CONSTANT FREQUENCY A-C POWER.

In addition to the a-c output of the engine driven generator, two inverters (No. 1 and No. 2) are provided to supply three phase, 115 volt, 400 cycle alternating current. Both inverters are powered by dc from the essential bus and each one is protected from overloads by a circuit breaker.

Note

On certain airplanes,[1] the inverters are called main (instead of No. 1) and stand-by (instead of No. 2) and the two positions of the inverter changeover selector switch are MAIN and STANDBY. Except for the differences in nomenclature, these inverter systems are identical.

Both inverters operate continuously. Normally, the No. 1 inverter powers the G-2 compass, gyro horizon, fuel flowmeter, fuel quantity indicators and, through a 26 volt transformer, the oil pressure indicator. The No. 2 inverter normally powers certain circuits of the AN/APG-30 system. A guarded inverter changeover selector switch and an inverter indicator light are provided on the right console. The inverter changeover selector switch has two positions; #1 INV. and #2 INV. When this switch is set to #1 INV., equipment normally powered by the No. 1 and No. 2 inverters, as described above, is energized. In the event of No. 1 inverter failure while the inverter changeover selector switch is set at #1 INV., the inverter indicator light will glow. Setting the inverter changeover selector switch to #2 INV. diverts the output of the No. 2 inverter to equipment normally powered by the No. 1 inverter. However, power is then no longer available to energize the AN/APG-30 circuits. In the event of No. 2 inverter failure while the inverter changeover selector switch is set at #2 INV., the inverter indicator light will glow.

Note

The inverter warning light will indicate only the failure of the No. 1 inverter when the inverter changeover selector switch is set at #1 INV. and the failure of the No. 2 inverter when the switch is set at #2 INV.

D-C POWER.

An engine driven combination a-c—d-c generator, rated at 200 amperes at a potential of 30 volts, is regulated to 27.7 +0.2 —0.2 volts and forms the principal direct current power source. It provides the required electrical energy to operate the various units of electrical equipment and, in addition, maintains the battery in a charged condition. A 24 volt, 36 ampere-hour storage battery serves as a stand-by or emergency source of power. An external source of direct current may be connected to the airplane for ground operation and

[1] F9F-6 airplanes ser No. 126257 through 127464.

test purposes. Power distribution is accomplished through a multiple bus network; i.e., battery, essential, main and monitor. The three position battery and generator switch controls system operation and provides manual override in emergencies to permit operation of certain equipment normally deenergized when generator failure occurs. The generator system is equipped with standard controls; i.e., generator control relay and voltage regulator. The battery relay, battery relay transfer switch, monitor bus relay "A", monitor bus relay "B", essential bus relay, ammeter shunt, voltammeter generator warning light and system external power receptacle complete the d-c power supply system.

The battery, generator and the external power source are connected in parallel with the main bus. The battery is connected through the battery relay, the generator through the generator control relay and external power through the system external power receptacle. The main bus, therefore, serves as the main distribution point for d-c power. The battery relay may be energized directly from the storage battery or from the generator, depending upon system requirements.

Monitor bus relay "A" and "B" are energized from the IND terminal of the generator control relay when the generated potential is above battery potential, or from the system external power receptacle when an external source of electrical power is connected to the airplane. Monitor bus relay "A" transfers generator or external power from the main bus to the essential and monitor busses. Monitor bus relay "B" functions to control the battery relay through the battery relay transfer switch when the landing gear control lever is set to UP. In this condition, the battery relay is energized by generator power received from the monitor bus. Since the main bus is connected to the battery bus through the battery relay contacts, the main bus and connecting circuits are automatically deenergized when generator operation is interrupted. Monitor bus relay "B" also serves to control the generator warning light circuit.

The essential bus relay transfers power from the battery bus to the essential bus. The essential bus relay is energized from either the battery bus or the main bus when the battery and generator switch is set to BAT. & GEN. position. The essential bus relay is energized by battery power on the battery bus, or by generator or external power on the main bus, providing this potential is greater than that of the battery.

EXTERNAL POWER RECEPTACLES.

A-C EXTERNAL POWER RECEPTACLE.

An external a-c power receptacle is provided in the nose wheel well for ground operation and testing of the AN/APG-30 gun ranging radar equipment and the AN/APX-6 IFF equipment.

D-C EXTERNAL POWER RECEPTACLES.[1]

Two external d-c power receptacles are provided on the underside of the fuselage aft of the main wheels. One receptacle is for engine starting only, the other supplies power to the entire d-c system with the exception of the starter. When external d-c power is plugged in for the d-c system, the battery, main, essential and monitor busses are automatically energized and all equipment normally powered by these busses may be operated.

D-C EXTERNAL POWER RECEPTACLES.[2]

This system is similar to that described in the above paragraph except that with external power plugged in and the battery and generator switch in the OFF position, only the main, essential and monitor busses are energized. The exclusion of the battery bus prevents the battery from being connected to the external source, eliminating the possibility of overcharging the battery.

NORMAL CONTROLS.

BATTERY SWITCH.

The single battery and generator toggle switch is on the electrical control panel, forward, on the right console (46, figure 1-3B). There are three switch settings: BAT. & GEN., OFF and BAT. The switch controls current to three bus bars, the main d-c bus, the essential bus and the monitor bus, which supply current to the various circuits in the airplane. With the landing gear control at DOWN (airplane in ground status), when the switch is set at BAT. & GEN., all three busses become energized and all electrical equipment may be operated (by operating the respective control switches), provided generator voltage is in excess of battery voltage or external power is plugged in. In flight status, should the generator fail, or if output is not in excess of battery voltage (indicated by warning light glow), the monitor and main d-c busses will be disconnected automatically and all circuits connected to these busses made inoperative. However, the essential bus will remain live and those circuits connected to it will be operative (warning lights, engine instruments, flight instruments, instrument lights, fuel selector, fuel system, tank dump, fuel shut-off, wing fuel transfer and fuel booster). If the switch is set at BAT. in this situation, the main d-c bus will be energized again and the circuits connected to it will be operative, but the monitor bus will remain dead and its circuits (gun camera, utility receptacle and gun heater) will remain inoperative. However, when the switch is set to BAT., it is *mandatory that circuits not necessary be disconnected* by setting applicable switches to OFF, since *battery power is adequate for only a short time.* The essential bus supplies all circuits necessary to flight and is energized in both BAT. & GEN. and BAT. switch positions. Setting the switch to OFF cuts off all equipment except the IFF detonator and canopy external control circuits.

[1] F9F-6 airplanes ser No. 126257 through 128098 and F9F-6P airplanes ser No. 127473 through 131255.

[2] F9F-6 airplanes ser No. 128099 and subsequent and F9F-6P airplanes ser No. 134446 and subsequent.

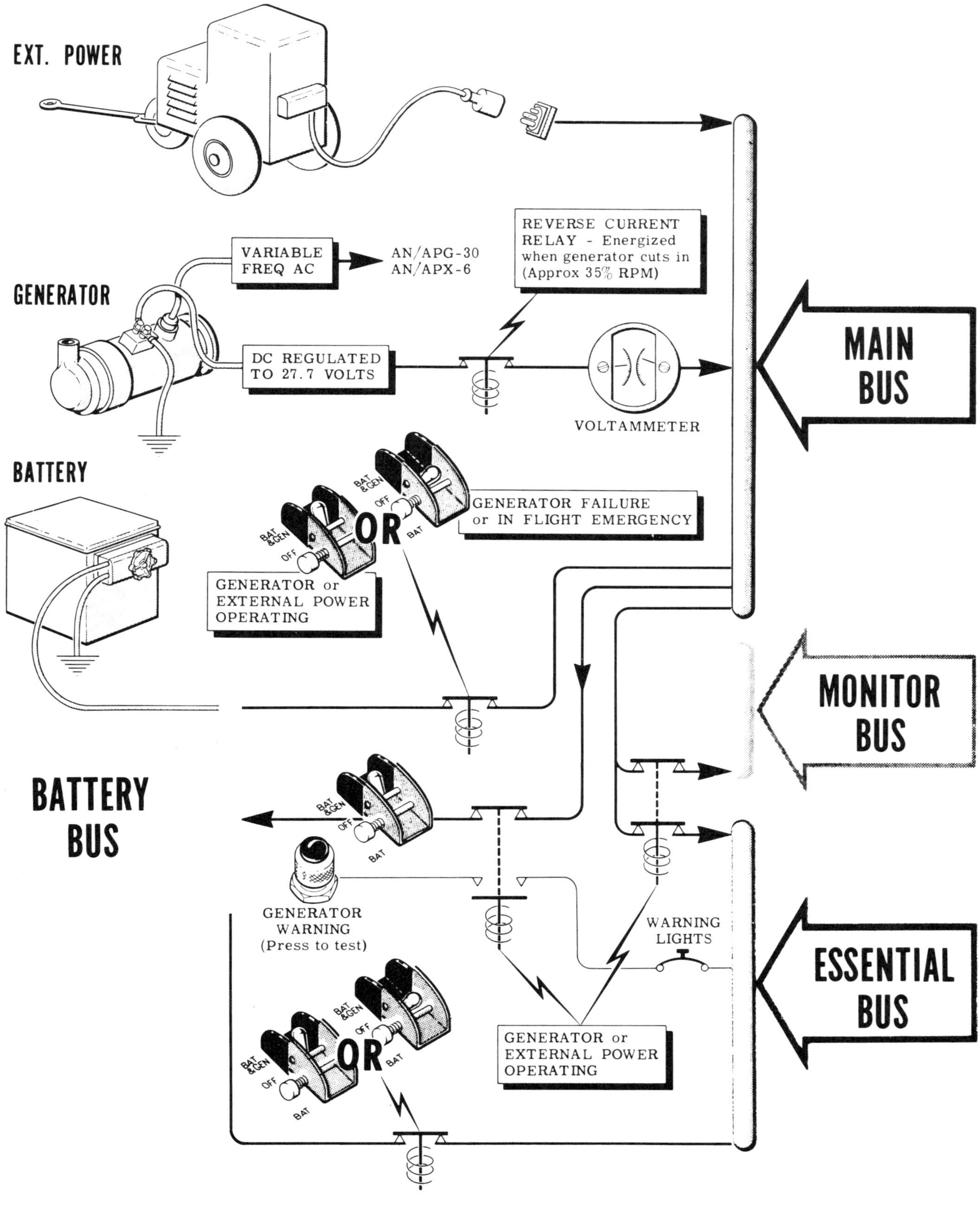

Figure 1-8. Electrical System Schematic Diagram (Sheet 1)

CIRCUIT BREAKER ON MAIN CIRCUIT BREAKER PANEL	EQUIPMENT POWERED THROUGH CIRCUIT BREAKER	APPROX CURRENT DRAIN (AMPERES)
MAIN BUS		
U. H. F. (See Note 5)	AN/ARC-27 Command Set Band Change Receiver Transmitter	 9.0 17.0 20.0
HYDRAULIC PUMP	Auxiliary Hydraulic Pump	37.6 (See Note 3)
STANDBY INVERTER	Stand-by Inverter	17.8
EXT. LIGHTS (2)	Approach Light Flasher Coder Wing Running Lights Wing Formation Lights Tail Lights Fuselage Lights Control Relays	0.8 (See Note 3) 0.3 1.5 0.3 0.8 5.4 0.2
A. F. C. S.	Armament Control System Gun Sight Recorder	4.5 0.2
PITOT (See Note 9)	Pitot Tube Heater	4.2
A. D. F.	AN/ARN-6 Radio Compass	4.0
RADAR APG-30	AN/APG-30 Radar Equipment	3.6
CABIN PRESS.	Air Conditioning System Main Shut-off Air Conditioning System Ram Air Valve Cabin Pressure Regulator Temperature Control Valve Temperature Control Box	0.3 1.1 0.5 1.1 0.2
ANT. RELAY (See Note 6)	AN/ARC-27, AN/ARR-2A Antenna Selector Relay	2.3 (See Note 3)
NAV.	AN/ARR-2A Radio Navigation Set	2.0
I. F. F. POWER	AN/APX-6 IFF Equipment	1.4
HOOK CONTROL	Arresting Hook Retracting Valve Arresting Hook Indicator Light Control Relays	0.1 (See Note 3) 0.2 (See Note 3) 0.9 (See Note 3)
(Armament) MASTER (See Note 8)	Armament Master Relay	0.6
GROUND LOCK	Ground Lock Solenoid	0.4
EMER. HYD. PUMP (See Note 7)	Auxiliary Hydraulic Pump Relay	0.4 (See Note 3)
DIVE BRAKE (See Note 10)	Speed Brake Hydraulic Valve	0.2 (See Note 3)
GUN CHARGING	Gun Charging Valve Hydraulic Pressure Switch and Control Relays	0.6 (See Note 3) 0.7 (See Note 3)
FLIGHT DEVICES	Speed Brake Limit Relay	0.1 (See Note 3)
MONITOR BUS		
GUN CAMERA	Gun Camera	1.6
UTILITY RECEPT.	—	—

CIRCUIT BREAKER ON MAIN CIRCUIT BREAKER PANEL	EQUIPMENT POWERED THROUGH CIRCUIT BREAKER	APPROX CURRENT DRAIN (AMPERES)
ESSENTIAL BUS		
FUEL BOOSTER	Fuel Boost Pump	30.5
EMER PWR (See Note 4)	Stabilizer Actuator and Control Relay Flying Tail Shut-off Valve	25.1 (See Note 3) 1.0 (See Note 3)
PRIM PWR (See Note 4)	Stabilizer Actuator	25.0 (See Note 3)
MAIN INVERTER	Main Inverter and Control Relays	20.3
FUEL TRANSFER	Fuel Transfer Pump	15.0
STARTING CONTROL	Ignition System and Control Relays	10.2 (See Note 3)
WING TANK DUMP	Wing Tank Dump Valves	7.5 (See Note 3)
PRIM CONT (See Note 4)	Stabilizer Up or Down Control Relays Wing Trimmer Actuator	4.6 (See Note 3) 1.1 (See Note 3)
WING FUEL SHUT-OFF	Wing Tank Fuel Shut-off Valves Fuel Transfer Pump Relay Wing Tank Empty Warning Lights	4.0 (See Note 3) 0.3 0.3 (See Note 3)
EMER CONT (See Note 4)	Stabilizer Up or Down Control Relays	4.6 (See Note 3)
FUEL SHUT-OFF	Main Fuel Shut-off Valve Fuel Transfer and Fuel Boost Pump Relays	2.0 (See Note 3) 0.7
FLIGHT INSTRS.	Wheels and Flaps Position Indicator Speed Brakes Position Indicator Stabilizer Position Indicator Rudder Trim Tab Position Indicator G-2 Compass	0.2 0.1 0.1 0.1 0.9
INSTR. LIGHTS	Instrument Panel Lights	1.4
COCKPIT LIGHTS	Cabin Lights Radio Panel Edge Lights Check-off List Lights Exterior Lights Panel Edge Lights	0.7 0.4 0.1 0.1
FLAPERETTE	Flaperette System By-pass Valve and Control Relay	1.1 (See Note 3)
WARNING LIGHTS	Inverter Indicator Light Generator Warning Light Fire Warning Lights	0.2 0.2 0.2
ENGINE INSTRS.	Tailpipe Temperature Indicator Low Fuel Level Warning Light	0.2 (See Note 3) 0.2 (See Note 3)
ENG. FUEL SYSTEM	Low Fuel Pressure Warning Light Main High Pressure Fuel Pumps Warning Light	0.2 (See Note 3) 0.2 (See Note 3)
EMER FUEL SYSTEM	Engine Fuel Control System Emergency Fuel System Indicator Light	0.2 (See Note 3) 0.2 (See Note 3)
BATTERY BUS		
IFF DET.	IFF Destructor	—

Note

1. System shown in operating condition.
2. Generator cuts in at 35% rpm; carries load at 46% rpm.
3. May be on during cruise condition.
4. Circuit breaker mounted on stabilizer power circuit breaker box on cabin after bulkhead.
5. AN/ARC-1 VHF command set with current drain of 10 amperes may be installed in place of AN/ARC-27.
6. Circuit breaker mounted on right console.
7. Switch type circuit breaker mounted on right console.
8. Switch type circuit breaker mounted on armament control panel on windshield deck. With switch set to ON and guns fired, current drain is approximately 102 amperes.
9. Switch type circuit breaker mounted on right console.
10. Circuit breaker mounted on left console.

Figure 1-8. Electrical System Schematic Diagram (Sheet 2)

CIRCUIT BREAKER RESET BUTTONS.

The reset buttons are grouped together on a panel on the inboard side of the right console (1, figure 1-3B). The button must be pushed in for operation of its respective circuit; if operation is desired without certain circuits, their applicable buttons may be pulled out.

INVERTER CHANGEOVER SWITCH.

This guarded switch (49, figure 1-3B), located on the right console inboard of the battery switch, has two positions: #1 INV. and #2 INV. Normally, the switch is placed at its guarded #1 INV. position, wherein the No. 1 inverter supplies power to the G-2 compass, gyro horizon indicator, fuel flowmeter and fuel quantity and oil pressure indicators. With the switch set at #1 INV., the No. 2 inverter powers certain circuits of the AN/APG-30 equipment. In the event of No. 1 inverter failure, as indicated by a glowing inverter warning light, raising the guard and setting the switch to #2 INV. automatically diverts the output of the No. 2 inverter to the equipment powered by the No. 1 inverter. This switch may also be used for testing the No. 2 inverter by setting the switch at #2 INV. Successful changeover from the No. 1 to the No. 2 inverter is indicated by the inverter warning light remaining out.

FUSES.

Fuses (three groups—DC INST, INTERIOR LTS and AC INST) are installed on the outboard edge of the right console (8, figure 1-3B). They are installed for additional protection to certain circuits. Failure of a unit will not blow a fuse if there is no circuit overload. Failure will cut out the portion of the circuit protected by fuses, and the remainder of the units on that circuit will remain operative.

SPARE FUSES.

Spare fuses are stored in a receptacle aft of the wing folding controls.

UTILITY RECEPTACLE.

A standard utility receptacle is on the right console aft (17, figure 1-3B).

EMERGENCY CONTROLS.

The battery and generator switch should be left at BAT. & GEN. position for emergency operation. To operate equipment connected to the main d-c bus, such as radio, it is necessary to set the switch to BAT. To conserve electrical power, *it is mandatory that all unnecessary equipment be turned off* when the switch is set to BAT.

INDICATORS.

VOLTAMMETER.

A dual dial, constant reading voltammeter is on the right console adjacent to the battery and generator switch. The meter indicates voltage at the main d-c bus and generator output amperage (50, figure 1-3B).

GENERATOR WARNING LIGHT.

A generator red warning light on the right console (47, figure 1-3B), adjacent to the battery and generator switch, will glow in the event of generator failure, or if output is below battery potential.

INVERTER WARNING LIGHT.

When the inverter switch is in its normal #1 INV. position, the red inverter warning light (48, figure 1-3B), on the right console, inboard of the battery switch will glow if the No. 1 inverter fails. When the inverter switch is set to #2 INV., the light will go out if this inverter is producing ac.

ELECTRICAL SYSTEM—F9F-6P AIRPLANES.

A-C POWER.

The a-c power supply circuit incorporates two inverters to deliver fixed frequency alternating current to the automatic pilot, Sonne camera, camera control, camera heating, G-2 compass and the fast erecting gyro circuits. Also supplied by the inverters are the AN/APX-6 IFF equipment, fuel flow, fuel quantity and oil pressure indicator circuits. The No. 2 inverter, replacing the main inverter of the F9F-6 airplanes, is the larger of the inverters. It is capable of producing 1500 volt-amperes at 3 phase, 400 cps output. The No. 1 inverter replaces the stand-by inverter and develops 250 volt-amperes at 3 phase, 400 cps output. In normal operation, with both inverters functioning, the No. 2 inverter output is exclusively providing all of the a-c power for the various systems. When the No. 1 inverter output is manually selected, it only provides power to the oil pressure indicator, G-2 compass, fast erecting gyro, fuel flow and the fuel quantity indicator circuits. The transition from the No. 2 inverter power supply to the No. 1 supply is accomplished by changeover relays within the circuit. A G-2 compass power adapter in the No. 2 inverter circuit corrects the sequence and also reduces the power output of the No. 2 inverter. A red inverter indication light is installed in the circuit and it will glow to indicate improper functioning of the selected inverter.

The circuit is placed in normal operation by setting the inverter changeover switch to the #2 INV. & AUTO PILOT position, which routes power from the inverter No. 2 circuit breaker to the remote control circuit breaker. This causes a relay within the circuit breaker to be energized which, in turn, completes the d-c circuit to the inverter, supplying the necessary d-c power for inverter operation. The remote control circuit breaker also provides thermal protection of the inverter power supply in abnormal conditions, such as overload of the circuit. In the #2 INV. & AUTO PILOT position, the changeover switch also routes power to the No. 1, 2, and 3 changeover relays and to the relay switch contact on the inverter, placing the inverter in operation. The energized No. 1, 2, and 3 relays supply the optional circuits with power from the No. 2 inverter. The circuits operated solely by power from the No. 2 inverter are supplied directly by the inverter. Inverter No. 2 power energizes the inverter warning light relay, causing the red inverter indicator light to be off. The light will continue to be off for the duration of inverter operation. The instrument transformer located in the circuit reduces the No. 2 inverter output to the 26 volts required by the oil pressure indicator circuit. The remaining transformer and the G-2 compass power adapter reduce the 200 volt output of the No. 2 inverter to the 115 volts required by the remaining circuits. Throughout the normal No. 2 inverter operation of the circuits, the No. 1 inverter is functioning, powered directly by direct current from the inverter No. 1 circuit breaker. The a-c output, however, is directed to ballast resistors through the changeover relay No. 3.

During the changeover operation of the circuit, effected by manually placing the inverter changeover switch in the #1 INV. position, the No. 2 inverter is inoperative and the No. 1, 2, and 3 relays are deenergized. The No. 1 inverter continues in operation but its output is diverted from the ballast resistor circuit and is routed through the deenergized relays to supply power to the G-2 compass, fast erecting gyro, oil pressure indicator, fuel flow and the fuel quantity indicator circuits. The inverter warning light relay is again energized.

Note

The continuous supply of power to the No. 1 inverter through the circuit breaker requires that the inverter No. 1 circuit breaker be pulled out in the event of malfunction or failure of the inverter, to prevent further damage to the inverter or other hazard.

D-C POWER.

The d-c power supply is the same as that in F9F-6 airplanes except that a d-c generator replaces the a-c—d-c generator.

HYDRAULIC POWER SUPPLY SYSTEM.

NORMAL CONTROLS AND OPERATION.

Fluid from the dual reservoir in the plenum chamber is pumped by a variable volume engine driven pump. The selector valve controls are in the cabin; most of the valves are outside the cabin and are operated through shafts and right angle drives. The power assist brake valves are on the brake pedals. The landing gear, wing folding, seat adjustment, canopy, gun charging, arresting hook and tail skid, and speed brakes selector valves are in the forward section of the fuselage; these last three are electrically controlled. In addition to the system reservoir, a small reservoir is provided for the wheel brakes. Pressure is distributed via two manifolds, a service manifold and a combat manifold; and a selector valve is installed, which may be set to shut off pressure to all systems not essential to flight, for combat operation. The auxiliary hydraulic pump is powered by the airplane electrical system. See figure 1-2 for hydraulic fluid specification and figure 1-8A for system schematic diagram.

Note

Hydraulic fuses, actuated by flow, are installed in the system to prevent excessive loss of fluid in case of failure of lines and/or units.
If a malfunctioning fuse should stop landing gear operation, the selector valve should be reversed momentarily to reset the fuse.

COMBAT SYSTEM SHUT-OFF CONTROL.

This control lever is on the right console, forward of the wing folding controls (11, figure 1-3B). It is moved outboard to HYDRAULIC PRESSURE OFF (combat operation), inboard to HYDRAULIC PRESSURE ON (normal operation—all hydraulic equipment operative). When the control is moved to HYDRAULIC PRESSURE OFF, only the following units are operative: speed brakes, gun chargers, seat adjustment, arresting hook, tail skid raising and flaperons. The following units are cut off when the control is moved to HYDRAULIC PRESSURE OFF to prevent loss of fluid due to damage to a non-essential system: canopy control, wing flaps, wing slats, wing folding, wheel brakes and landing gear.

AUXILIARY HYDRAULIC PUMP SWITCH.

This toggle switch, equipped with a guard, is on the right console aft, adjacent to the wing folding controls (18, figure 1-3B). It is set outboard to ON and inboard to OFF. If there is no action when some part of the system is operated, or if no pressure is indicated on the main system gage, the switch should be set to ON.

CAUTION

Do not use auxiliary pump on the ground without connecting external d-c power.

Do not leave auxiliary pump switch at ON during flight.

Note

Auxiliary pump operation of any system will be considerably slower than normal operation.

INDICATORS.

SYSTEM PRESSURE GAGE.

This gage is on the right console aft, adjacent to the map case (21, figure 1-3B). The dial shows pressure in psi, and is numbered from 0 to 2000. It should show 1500 psi when the engine is running and no hydraulic units are operating. Pressure indication may drop momentarily considerably below this figure when a system or systems are being operated.

AUXILIARY PUMP PRESSURE GAGE.

This gage is on the right console, aft of the system gage and forward of the auxiliary pump switch (20, figure 1-3B). The dial shows pressure in psi and is numbered from 0 to 2000. It should indicate approximately 1700 psi with no hydraulic device operating but may show

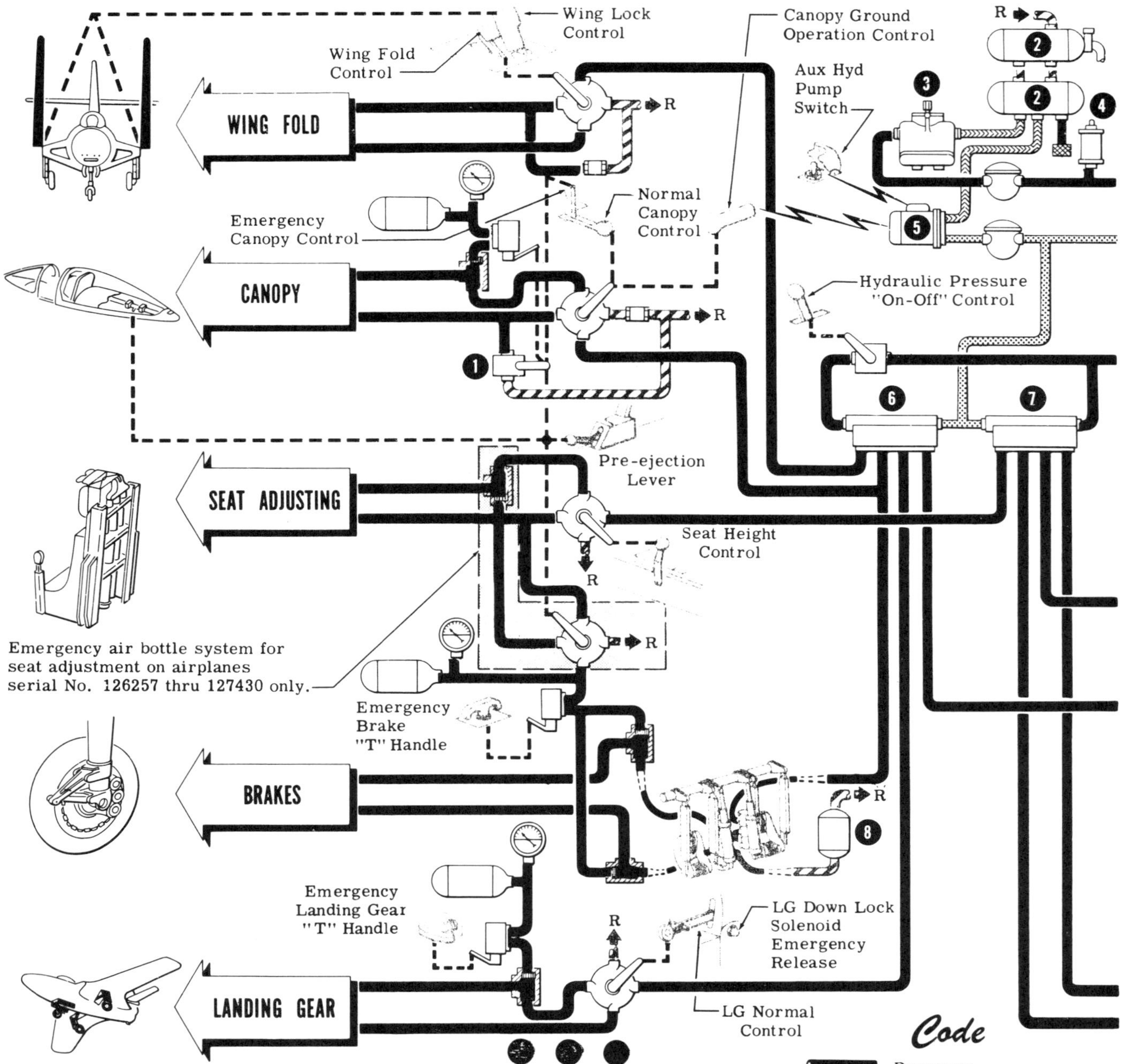

1. Canopy Dump Valve
2. Reservoir (Top and Bottom)
3. Engine-driven Main Hydraulic Pump
4. Main System Accumulator
5. Electric Auxiliary Hydraulic Pump
6. Service Manifold
7. Combat Manifold
8. Brake Reservoir
9. Flight Manifold
10. DELETED
11. Flaperon Hydraulic Pressure Sensing Switch
12. Vent and Relief Valve
13. Flaperette System Reservoir
14. Flaperette Pump (See Note)
15. Flaperette Accumulator
16. Relief Valve
17. Pressure Switch

Note

On airplanes equipped with J48-P-8 engines, the electric flaperette system hydraulic pump is replaced by an engine driven hydraulic pump. In addition to the engine driven pump, a normally closed by-pass valve is installed between the pressure and return lines. See paragraph Emergency Lateral Control System-Flaperettes.

Figure 1-8A. Hydraulic System Schematic Diagram (Sheet 1)

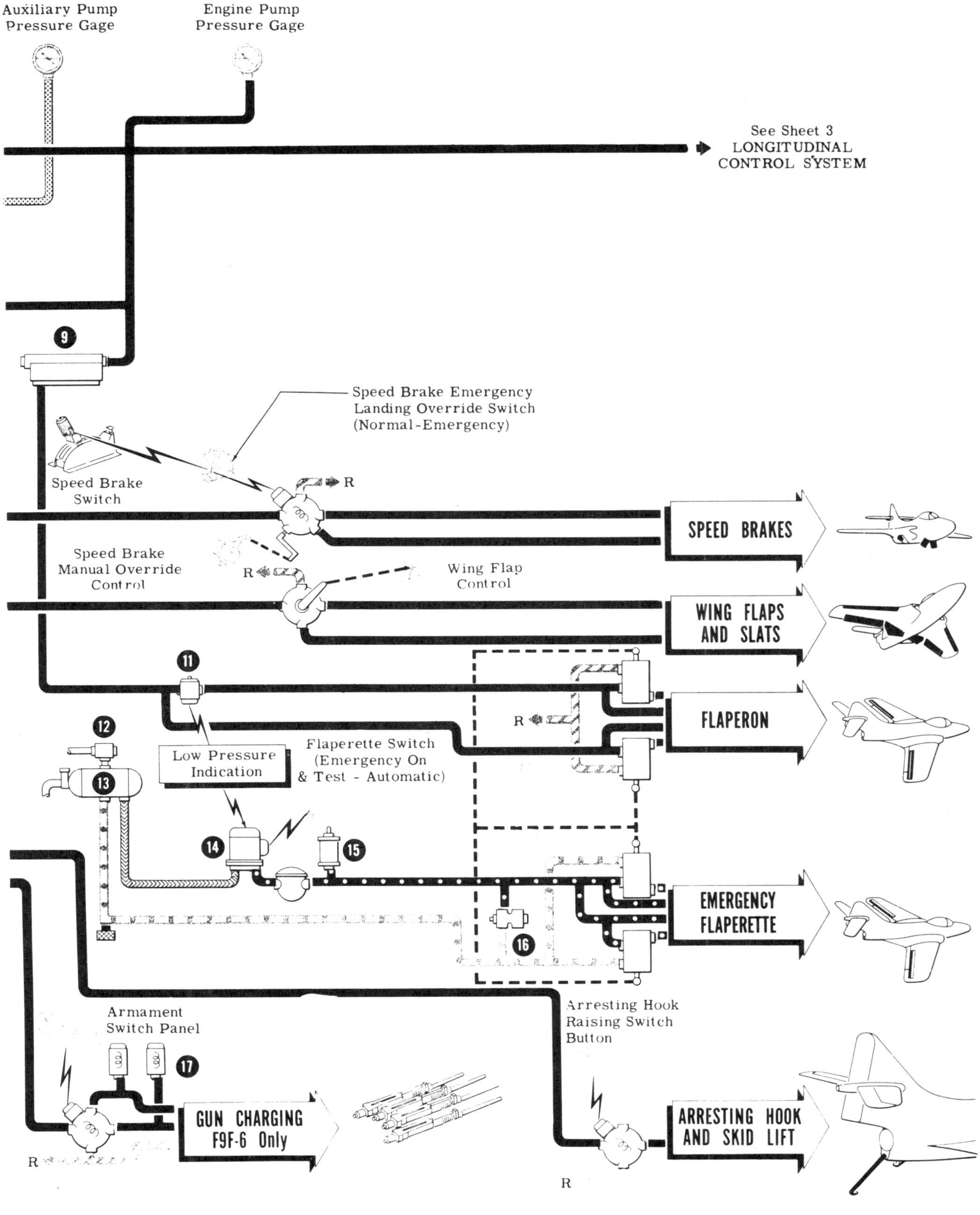

Figure 1-8A. Hydraulic System Schematic Diagram (Sheet 2)

INTERIM LONGITUDINAL CONTROL CONFIGURATION

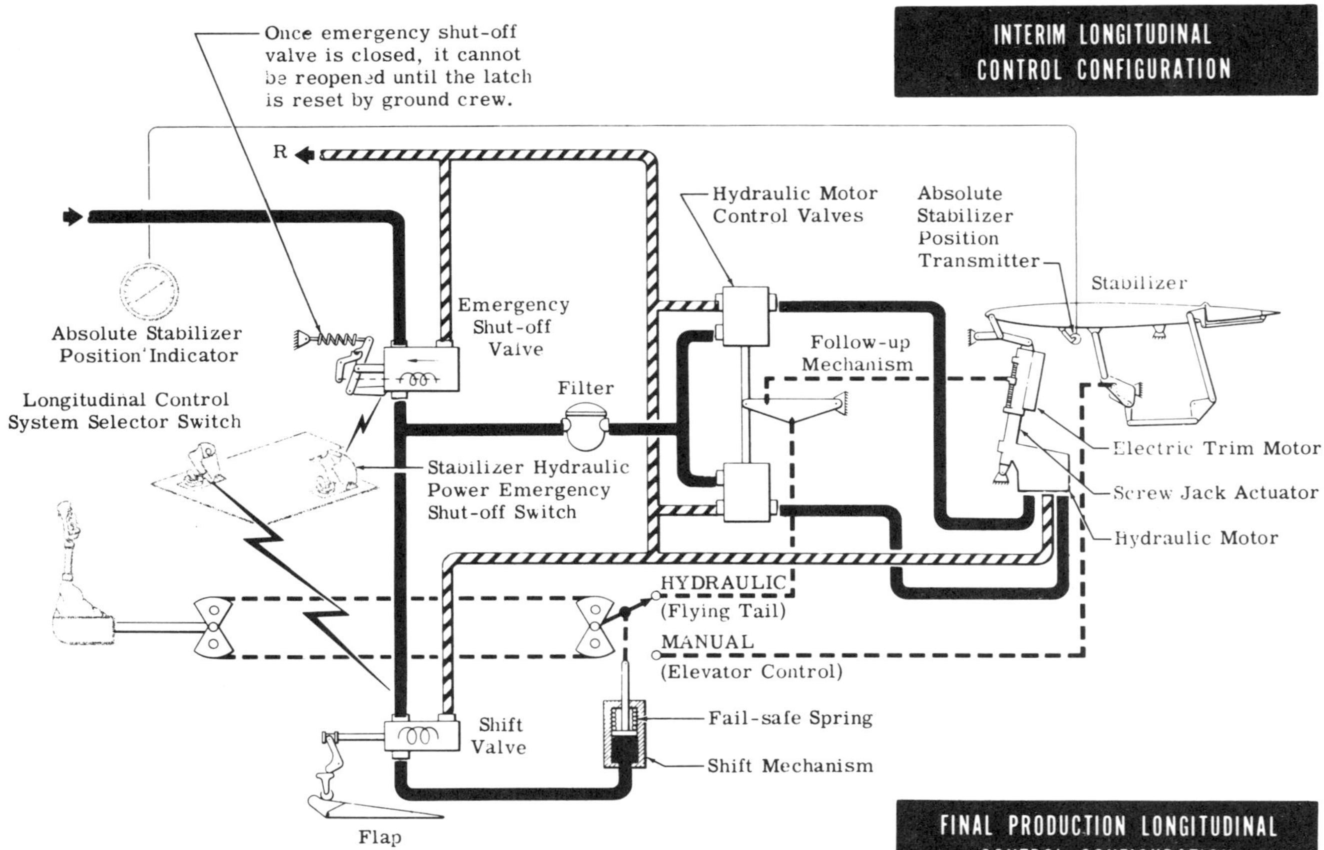

FINAL PRODUCTION LONGITUDINAL CONTROL CONFIGURATION

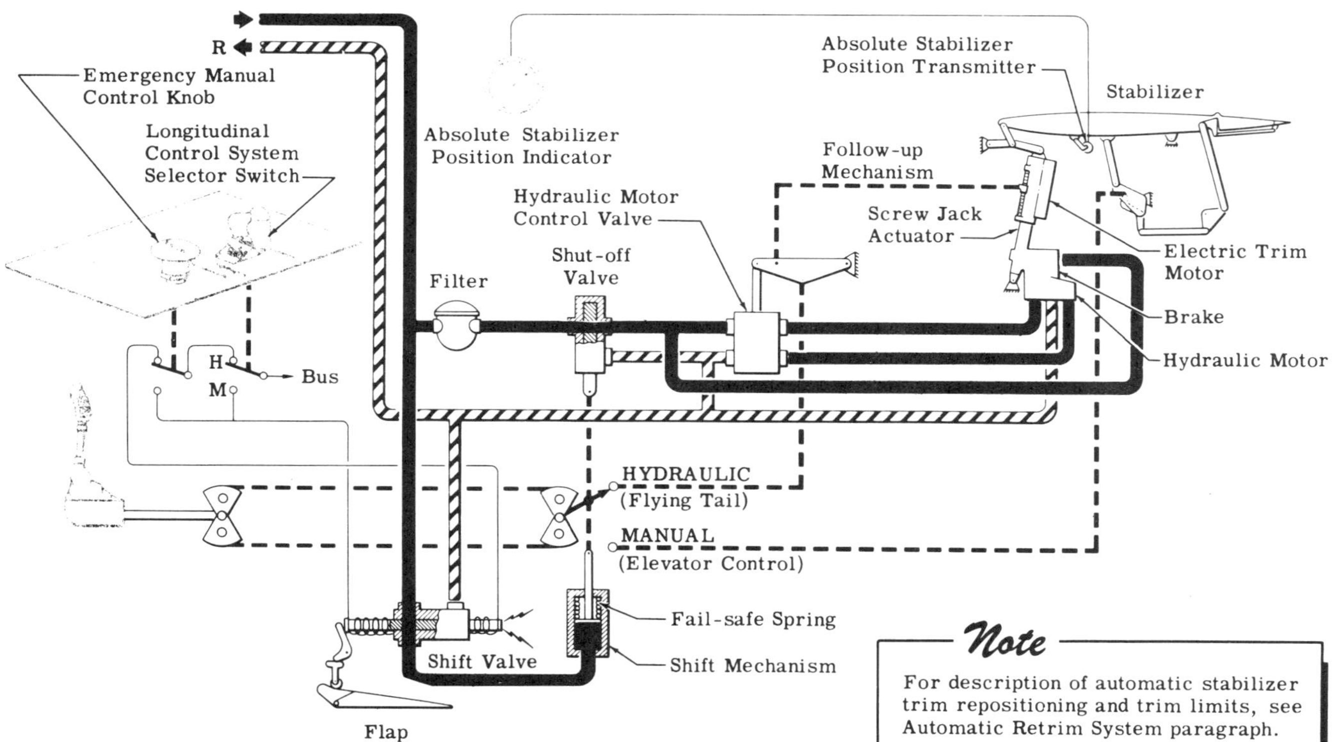

Note

For description of automatic stabilizer trim repositioning and trim limits, see Automatic Retrim System paragraph.

Figure 1-8A. Hydraulic System Schematic Diagram (Sheet 3)

very low readings, depending on the number of units being operated by the pump.

FLIGHT CONTROL SYSTEM.

This airplane is controlled in the conventional manner, using a control stick and pedals. Lateral and longitudinal control is achieved by the use of flaperons and the flying tail or conventional spring tab elevators. Flaps and slats are used to facilitate low speed flight. Speed brakes are provided to decelerate the airplane at any airspeed. In addition, an adjustable stabilizer and a wing trimmer are installed for trim.

NORMAL LATERAL CONTROL SYSTEM—FLAPERONS.

The control stick in the cabin is linked hydraulically to the flaperons which provide lateral control. For normal operation, lateral movement of the stick raises a flaperon from the wing surface. Hydraulic pressure to the flaperon valves and operating cylinders is derived from a direct line to the engine driven hydraulic pump.

EMERGENCY LATERAL CONTROL SYSTEM—FLAPERETTES.

The aft halves of the flaperons are flaperettes and provide a reduced amount of lateral control when the flaperon system is inoperative. Normally, the flaperettes are locked in line with the flaperons and move as part of them. In the event of a hydraulic failure, the changeover to the flaperette system is automatic. On airplanes equipped with J48-P-6A engines, this emergency system is essentially an independent hydraulic system which duplicates the operation of the main hydraulic system and includes an electrically driven hydraulic pump, reservoir, accumulator, operating valves and cylinders. On airplanes equipped with J48-P-8 engines, this emergency system differs from the system described above in that an additional engine driven hydraulic pump replaces the electric hydraulic pump and a by-pass valve is incorporated in the system.

FLAPERETTE SYSTEM—ELECTRICAL HYDRAULIC PUMP.

As long as hydraulic pressure is available for the flaperon system, a main hydraulic pressure sensing switch remains open and the electric hydraulic pump is deenergized. When main hydraulic system pressure drops, the pressure sensing switch closes to energize the electric hydraulic pump which provides hydraulic pressure to operate the flaperettes. If automatic changeover does not occur because of a malfunctioning pressure sensing switch, the electric hydraulic pump can be energized by setting the flaperette system control switch to EMER. ON & TEST.

FLAPERETTE SYSTEM—ENGINE DRIVEN HYDRAULIC PUMP.

As long as hydraulic pressure is available for the flaperon system, the by-pass valve is energized to remain open to by-pass flaperette pump pressure to the flaperette system reservoir. The by-pass valve is deenergized either by the opening of a main hydraulic system pressure sensing switch due to main system hydraulic pressure failure, or by setting the flaperette system control switch to EMER. ON & TEST. When the by-pass valve is deenergized, it closes, thus diverting hydraulic pressure to operate the flaperettes.

FLAPERETTE SYSTEM CONTROL SWITCH.

A two position guarded flaperette system switch is located on the right console (22, figure 1-3B). Normally, the switch is left in the AUTO position, wherein lateral control will automatically change over to the flaperette system in the event of main hydraulic system failure. Moving the switch to its other position (EMER. ON & TEST) serves two purposes. Primarily, it permits changing the lateral control over to the flaperette system in the event that automatic changeover does not occur due to the failure of the main hydraulic system pressure sensing switch. Secondly, it permits the flaperette system to be checked for proper operation.

EMERGENCY FLAPERETTE CONTROL SYSTEM—AIRPLANES WITH ENGINE DRIVEN FLAPERETTE HYDRAULIC PUMPS.

The emergency flaperette control system provides the pilot with an increase in the number of flaperette operating cycles available during landing approach under flame-out conditions. The system is installed because the engine windmilling rpm at low airspeeds is too low to provide sufficient hydraulic pressure for proper control. This system includes an air bottle, an additional hydraulic accumulator, a check valve, an emergency flaperette power control handle in the cabin, an air pressure gage in the cabin and interconnecting tubing.

EMERGENCY FLAPERETTE POWER CONTROL.

This control is located on the left console and is actuated in a slot which has ON and OFF detents. Setting the control to ON opens a valve which releases air pressure to the accumulator. Pressurized hydraulic fluid in the accumulator is directed to the flaperette system operating valves and cylinders. The number of flaperette operating cycles is approximately 22. The control valve may be set to OFF at any time to shut off air pressure to the accumulator.

AIR PRESSURE GAGE.

This gage (7, figure 1-3), located at the aft part of the left console, is calibrated in the number of flaperette operating cycles available. When the air bottle is fully charged, the gage will normally indicate more than 22 operating cycles even though the system is off. When the system is turned on and becomes pressurized, the gage needle will fall off to indicate that approximately 22 operating cycles are available. As the flaperette system is operated, the needle will move towards zero to show the number of available cycles remaining.

ARTIFICIAL FEEL.

Lateral stick forces are simulated by a cam-driven, spring loaded mechanism. The artificial feel thus induced in the lateral control system is necessary, since hydraulic operation of the flaperons and flaperettes prevents feedback of air loads to the control stick. To replace normal air load "feel", a cam attached to the artificial feel sector moves against a spring loaded arm, inducing a force in the system which opposes stick movement. This lateral force is independent of airspeed and is the same when either the normal flaperon system or the emergency flaperette system is being operated. Artificial feel is light and amounts to 16 pounds for full stick travel.

WING TRIMMER.

The wing trimmer is a movable surface on the left wing tip provided to control lateral out-of-rig conditions. It is operated electrically by an actuator in the wing and controlled by left and right movement of the spring loaded thumb switch on the control stick (35, figure 1-3A).

Note

Fore and aft movement of this switch is used for stabilizer trim.

Holding the switch to the left causes the wing trimmer surface to rise and the left wing to lower. Movement to the right causes the trimmer to lower and the left wing to rise. The switch is spring loaded to return to the central (off) position.

LONGITUDINAL CONTROL SYSTEM—INTERIM MODIFICATION.[1]

Note

The interim longitudinal control system modification may be installed on these airplanes by G.A.E.C. Service Bulletin No. 224. Until this interim or the final production modification is incorporated, all these airplanes are controlled by a conventional geared tab elevator control system. For airplanes equipped with the final production longitudinal control system modification, see Longitudinal Control System Final Production Modification paragraph.

DESCRIPTION.

Longitudinal control is provided by two separate systems. These are: the hydraulic pressure operated, all movable horizontal stabilizer control system for high speed clean configuration flight; and the manual elevator control system for low speed flight and emergency high speed flight. The shift from one system to the other is accomplished by one of two (normal and emergency) switches. An absolute stabilizer position indicator provides a means of checking stabilizer movement and position at all times.

FLYING TAIL (HYDRAULIC POWER OPERATED ALL MOVABLE HORIZONTAL STABILIZER) CONTROL SYSTEM.

The powered longitudinal control system (flying tail) may be activated in one of two ways: first, in the clean configuration (flaps up), by moving the longitudinal control system selector switch to HYDRAULIC; and second, by raising the wing flaps, with the longitudinal control system selector switch set to HYDRAULIC. In both cases, hydraulic pressure is directed to the shift valve to transfer the longitudinal control system to the hydraulic power operated, all movable horizontal stabilizer, with leading or helping elevators. In this configuration, moving the control stick mechanically actuates valves which direct hydraulic pressure to a hydraulic motor. The motor rotates a screw jack which moves the leading edge of the stabilizer up or down. As the stabilizer pivots, a mechanical linkage deflects the elevators to provide additional control effectiveness. Thus, moving the control stick aft lowers the leading edge of the stabilizer and raises the trailing edge of the elevators. Moving the control stick forward raises the leading edge of the stabilizer and lowers the trailing edge of the elevators. An absolute stabilizer position indicator is mounted on the upper left side of the instrument panel.

A follow-up mechanism is provided to stop the hydraulic motor when the stabilizer position corresponds to the control stick position. Since there is no aerodynamic force feedback to the control stick, an artificial feel system is installed. This artificial feel consists of an initial torsion bar spring which provides stick centering near neutral, and a second preloaded feel spring which, with the bob weight, provides realistic control force for stability and maneuvering.

The stabilizer hydraulic power emergency shut-off switch, located on the inboard side of the stabilizer control selection panel on the left console, operates a shut-off valve between the pressure source and the hydraulic components of flying tail. When the shut-off valve is closed, it prevents further pressure from reaching the flying tail components, and at the same time, transfers longitudinal control to manual operation.

NORMAL CONTROL.

LONGITUDINAL CONTROL SYSTEM SELECTOR SWITCH.

The longitudinal control system selector switch, located on the outboard side of the stabilizer control selection panel on the left console (36, figure 1-2A), controls the operation of the shift valve. When the switch is in the HYDRAULIC position, the shift valve permits hydraulic pressure to act upon the shift actuator mounted on the aft control sector. The hydraulic control rod is attached to this sector and linked to the stabilizer control valves. The position of the valve pistons is con-

[1] F9F-6 airplanes ser No. 126257 through 127335, 130977 through 131036 and F9F-6P airplanes ser No. 127473 through 134457.

trolled by movement of the control stick through the control rod linkage. Hydraulic pressure is transmitted through the valves to a hydraulic motor, which operates the screw jack stabilizer actuator. During hydraulic operation, both the stabilizer and elevators move in response to control stick movement. When the switch is set to the MANUAL position, the shift valve is closed and no pressure is exerted upon the shift actuator. The manual control rod attached to the aft control sector, being spring loaded, returns to the operative position. During manual operation, the stabilizer remains stationary and the elevators move in response to control stick movement.

EMERGENCY CONTROL.

STABILIZER HYDRAULIC POWER EMERGENCY SHUT-OFF SWITCH.

The guarded emergency shut-off switch is mounted on the inboard side of the stabilizer control selection panel on the left console (38, figure 1-2A). The switch normally remains in the guarded position. If there is a malfunction of the hydraulic components of the flying tail, the switch is placed in the emergency position. Actuating this switch closes the shut-off valve, which shuts off hydraulic pressure to the stabilizer control valves and the shift valve. The stabilizer control system will then automatically shift to manual operation.

CAUTION

> If, during flight, the shut-off switch has been actuated, the flying tail control mechanism must be completely checked to determine the reason for the malfunction. Any improper condition of the mechanism must be corrected and the solenoid latch on the shut-off valve must be reset by the ground crew before making the next flight in the airplane.

MANUAL ELEVATOR CONTROL SYSTEM.

Lowering the wing flaps actuates a valve to relieve hydraulic pressure in the stabilizer control system shift actuator. This permits the shift actuator spring to change control automatically from the hydraulic power operated, all movable tail configuration to the manual elevator control configuration. In addition, when the flaps are lowered, the balancing tab neutral position is repositioned to 8-3/4 degrees down and a system bungee is automatically connected into the manual control system to provide satisfactory trim change and to improve airplane stability in the flaps down configuration.

Note

> At airspeeds over 150 knots, if the speed brakes are kept extended when the flaps are lowered (by setting the speed brakes emergency landing override switch to its emergency position), automatic shifting from powered longitudinal to manual elevator control will not occur. Shifting will not occur, because the increased air loads upon the flaps with speed brakes extended, prevent the flaps from lowering sufficiently to actuate the shift valve.

Longitudinal trimming by means of the stabilizer is performed in the same manner as with the powered control system, except that the control stick trim position varies with airspeed.

EMERGENCY MANUAL ELEVATOR CONTROL.

In the event of mechanical failure while flying in the hydraulic power operated, all movable horizontal stabilizer configuration, placement of the longitudinal control system selector switch to MANUAL, relieves hydraulic pressure in the stabilizer control system shift actuator. This permits the shift actuator spring to change control automatically from the hydraulic power operated, all movable horizontal stabilizer configuration to the manual elevator control configuration.

ABSOLUTE STABILIZER POSITION INDICATOR.

This indicator is mounted on the upper left side of the instrument panel. It is calibrated in degrees to indicate absolute stabilizer position, as determined by control stick movement. The range of the indicator is 0 to 4-1/2 degrees airplane nose-down, and 0 to 6-1/2 degrees airplane nose-up. On the manual control system, the indicator always denotes the trim position of the stabilizer and will move only when the stabilizer trim setting is changed. When operating on the hydraulic powered control system, the indicator continuously denotes the true stabilizer position and moves not only when the stabilizer trim adjustment is changed, but also whenever the control stick is displaced from neutral. Thus, when operating on the hydraulic powered control system, IT IS IMPORTANT TO REALIZE THAT THE ABSOLUTE STABILIZER POSITION INDICATOR DENOTES TRIM POSITION ONLY WHEN THE CONTROL STICK IS IN NEUTRAL AND NO FORE OR AFT CONTROL STICK FORCE IS BEING APPLIED.

WARNING

> Prior to take-off, shift longitudinal control system to manual control and adjust stabilizer (3-1/2 degrees airplane nose-up trim).

ADJUSTABLE STABILIZER.

Longitudinal trim is obtained through movement of the entire horizontal tail surface by an electrically operated stabilizer actuator mounted in the tail section. Limit switches in the stabilizer actuator restrict the stabilizer within the limits of its electrical adjustment range. D-C power to operate the electric motor is taken off the essential bus, which remains energized despite generator failure. Normally, stabilizer adjustment is controlled by a momentary type thumb switch on the control stick grip. An emergency stabilizer adjustment control system is

STABILIZER POSITION vs CONTROL STICK POSITION FOR VARIOUS STABILIZER TRIM SETTINGS

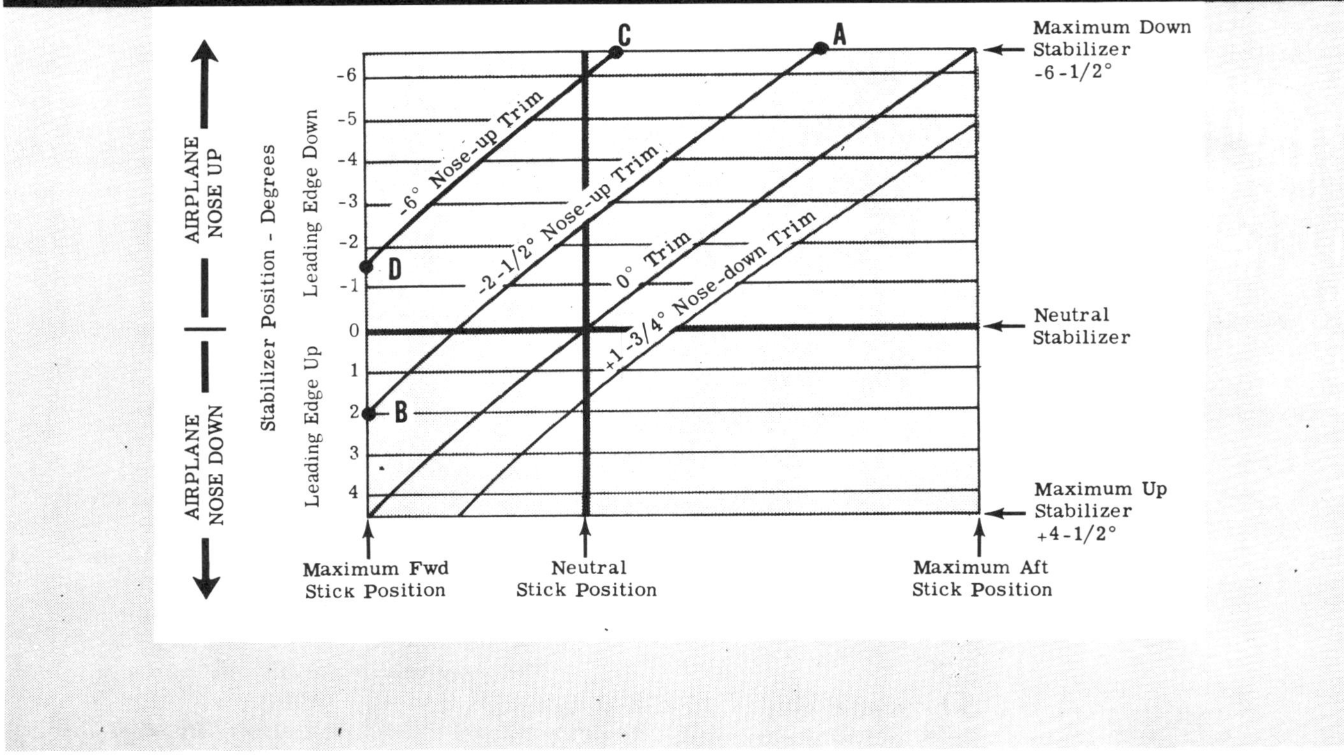

Figure 1-8B. Stabilizer Position vs Control Stick Position

provided, which includes a duplicate electric circuit and two control switches. A stabilizer trim position indicator is mounted on the right console.

ADJUSTABLE STABILIZER NORMAL CONTROL.

The system is normally operated by the momentary thumb switch on top of the control stick grip (16, figure 1-4). It is moved forward for nose-down trim, aft for nose-up trim, and is spring loaded to return to the central (off) position. Stabilizer position in degrees is indicated by a stabilizer position indicator.

Note

Lateral movement of this switch controls the wing trimmer.

The range of stabilizer travel available is from 6 degrees airplane nose-up trim to 1-3/4 degrees airplane nose-down trim.

Note

It is necessary that the stabilizer adjustment emergency power switch be set at NORM. in order to energize the normal stabilizer adjustment control system.

ADJUSTABLE STABILIZER EMERGENCY CONTROL SYSTEM.

In the event of an open or short circuit in the normal wiring system, the stabilizer may be adjusted by means of two switches on the left console. These switches control power from the essential bus to the electric motor through a duplicate electric circuit.

ADJUSTABLE STABILIZER EMERGENCY CONTROLS.

EMERGENCY ELECTRIC POWER CONTROL SWITCH.

This guarded switch, mounted on the stabilizer power panel on the left console (25, figure 1-2A), has three positions: NORM., OFF, and EMER. In the NORM. position, the stabilizer is adjusted by the momentary thumb switch on the control stick grip. In the OFF position, neither the normal or emergency systems can be operated. In the EMER. position, the stabilizer is adjusted by the stabilizer emergency control switch.

STABILIZER EMERGENCY CONTROL SWITCH.

This two position momentary type toggle switch, mounted on the stabilizer power panel on the left console (24, figure 1-2A), has two positions: forward, NOSE DOWN; aft, NOSE UP. Holding the switch in either of these two positions permits adjustment of the stabilizer to obtain desired airplane trim.

Note

The emergency electric power control switch must be set at EMER. to adjust the stabilizer with the stabilizer emergency control switch.

Stabilizer position in degrees is indicated by the stabilizer position indicator.

STABILIZER POSITION INDICATOR.

This indicator is mounted on the left console (21, figure 1-2A). It is calibrated in degrees to indicate stabilizer position. The range of the indicator is 0 to 2-1/2 degrees airplane nose-down trim and 0 to 6-1/2 degrees airplane nose-up trim.

Note

When the flying tail modification is installed, the stabilizer position indicator will be replaced by an absolute stabilizer position indicator mounted on the instrument panel. See Absolute Stabilizer Position Indicator paragraph.

LONGITUDINAL CONTROL SYSTEM—FINAL PRODUCTION MODIFICATION.[1]

Note

The final production longitudinal control system may be installed on prior airplanes by F9F Aircraft Service Change No. 256.

DESCRIPTION.

Longitudinal control is provided by two separate systems. These are: the hydraulic pressure operated, all movable horizontal stabilizer control system for high speed clean configuration flight; and the manual elevator control system for low speed flight and emergency high speed flight. The shift from one system to the other is accomplished by one of two (normal and emergency) switches. An absolute stabilizer position indicator provides a means of checking stabilizer movement and position at all times.

FLYING TAIL (HYDRAULIC POWER OPERATED ALL MOVABLE HORIZONTAL STABILIZER) CONTROL SYSTEM.

The hydraulic pressure to operate the powered longitudinal control system comes from an engine driven pump. (See the Hydraulic Power Supply System paragraph.) The d-c power necessary to shift from one longitudinal control system to the other comes from the essential bus. The flying tail shift circuit breaker (43, figure 1-3), mounted on the air conditioning panel, protects the longitudinal control system selector circuit. The powered longitudinal control system may be activated in one of two ways; by moving the longitudinal control system selector switch to HYDRAULIC when the wing flaps are up (clean configuration), or by raising the wing flaps when the longitudinal control system selector switch has already been set to HYDRAULIC. In both cases longitudinal control is transferred from a trimming stabilizer and movable elevator to a hydraulically operated flying tail. The flying tail is moved up or down by the stick as it operates a valve which controls extension or retraction of an irreversible screw jack attached to the stabilizer. Artificial feel in incorporated because there is no aerodynamic feedback in an irreversible control system. On the powered longitudinal control system, the elevators are geared to move slightly at a fixed ratio with the stabilizer. Their rigging is such that under all level flight conditions the elevators trail up.

FLYING TAIL SYSTEM OPERATING CONDITIONS.

Movement of the stick fore and aft of neutral moves the stabilizer leading edge up and down, respectively, by causing the (screw jack) actuator to extend and retract hydraulically. Electrical operation of the actuator establishes the trim position of the stabilizer. The stabilizer will be at its trim position only when the stick is in neutral, since any displacement of the stick out of neutral will also move the stabilizer hydraulically.

The maximum stabilizer deflection positions of 6-1/2 degree down and 4-1/2 degrees up give a total travel of 11 degrees. This full stabilizer travel, however, is only possible at zero trim. At any other trim setting, the total travel for full stick throw will be less than 11 degrees by the amount of trim selected. Regardless of the trim setting, full aft stick throw from neutral will not produce a stabilizer down travel of more than 6-1/2 degrees, and full forward stick throw from neutral will not produce a stabilizer up travel of more than 4-1/2 degrees. The effect of trim setting on the total stabilizer travel when using the power control system is shown in the chart of figure 1-8B.

The following example will outline the use of the chart to show the effect of various trim settings on the stabilizer travel for full stick throw. Assume that the stabilizer has been trimmed to 2-1/2 degrees airplane nose-up. The stabilizer setting is therefore −2-1/2 degrees. Locate the −2-1/2 degree nose-up trim line on the chart. Follow the line to the right, in the direction of aft stick motion, until point A is reached. Note that point A lies on the −6-1/2 degree stabilizer position line and is between the neutral and aft stick stop positions. This means that with 2-1/2 degrees nose-up trim, the maximum stabilizer down travel of 6-1/2 degrees is reached before the stick has bottomed on its aft stop. The stick can be moved aft from point A until it bottoms on its stop but further motion of the stabilizer will not occur because the actuator has bottomed against its internal stops. Motion of the stick aft of point A, however, will cause the hydraulic servo valve to bottom in its open po-

[1] F9F-6 airplanes ser No. 127336 through 127356, 131037 and subsequent and F9F-6P airplanes ser No. 134458 and subsequent.

sition. If the servo valve bottoms before the stick reaches its aft stop, a flexible safety link will absorb the excess motion, thus protecting the servo valve from damage. Compression of the safety link will be felt as an increase in stick load. Now, follow the −2-1/2 degree nose-up trim line to the left, in the direction of forward stick motion, until point B is reached. Note that point B is at the intersection of a line representing the forward stick stop and a line representing the +2 degree stabilizer position. This means that the stick has been moved as far forward as possible, but that the maximum leading edge up position of the stabilizer attained with the −2-1/2 degree trim position is just +2 degrees, or a reduction in the stabilizer up travel from zero degrees by an amount equal to the stabilizer leading edge down trim (4-1/2 degrees maximum stabilizer leading edge up minus 2-1/2 degrees stabilizer leading edge down trim equals 2 degrees actual stabilizer leading edge up position for full forward stick travel). Note also that at the neutral stick position, the −2-1/2 degree nose-up trim line intersects the −2-1/2 degree stabilizer position line, confirming the fact that the stabilizer will be at the trim position selected when the stick is neutral for the power control system.

The chart also illustrates that large amounts of trim severely limit the total stabilizer travel. Locate the −6 degree nose-up trim line and follow it to its extremities, represented by points C and D. Point D is reached when the stick is against its forward stick stop, but point C is reached when the stick is only a short distance aft of the neutral stick position. Point D represents a stabilizer position of −1-1/2 degrees and point C represents a stabilizer position of −6-1/2 degrees, or a total stabilizer travel of 5 degrees as compared to 11 degrees for zero trim.

MANUAL ELEVATOR CONTROL SYSTEM.

The manual control system consists of the electrically operated stabilizer and standard mechanically controlled elevators. The airplane may, under normal conditions, be taken out of the powered longitudinal control system by moving the longitudinal control system selector switch to MANUAL, while remaining in the clean configuration (flaps up), or by lowering the flaps even though the longitudinal control system selector switch is at HYDRAULIC. The longitudinal control system selector switch controls the shift by actuating a valve that controls the hydraulic pressure acting upon the shift mechanism. Actuating the flaps controls the same valve through mechanical linkage.

Note

At airspeeds over 150 knots, if the speed brakes are kept extended when the flaps are lowered (by setting the speed brakes emergency landing override switch to its emergency position), automatic shifting from powered longitudinal to manual elevator control will not occur. Shifting will not occur, because the increased air loads upon the flaps with speed brakes extended, prevent the flaps from lowering sufficiently to actuate the shift valve.

LONGITUDINAL CONTROL SYSTEM SHIFT.

The shift from one system to the other is accomplished by the shift mechanism on the aft control sector. The shift from manual power control occurs in approximately 2-1/2 seconds. Hydraulic pressure holds the shift mechanism in the powered longitudinal control position by overcoming a fail-safe spring, preloaded to force the shift mechanism to manual elevator control. Thus, when the longitudinal control system selector switch is set to MANUAL, or if there is a loss of hydraulic pressure, the spring will cause a shift to manual elevator control. The shift from hydraulic control to manual control occurs in approximately 1 to 1-1/2 seconds. During shift-over, both hydraulic and manual control systems are partially effective and there is no loss of control. The two control systems are adjusted with respect to each other so that when shifting from the powered system to the manual system in straight 1.0g trimmed flight (level, climb or glide), the nose of the airplane will rise slightly. If the shift is made during a turn or "g" maneuver, trim change during the shift will be small but stick forces required to hold the same "g" will more than double. If the shift is made during a turn or "g" maneuver at very high Mach numbers, trim change will be low but electric stabilizer trim should be used as necessary to control the flight path of the airplane in recovering to level flight.

LONGITUDINAL CONTROL SYSTEM SELECTOR SWITCH.

The guarded longitudinal control system selector switch (44, figure 1-3) is located on the inboard side of the stabilizer control selection panel on the left console. The two positions of this toggle switch enable the pilot to select either longitudinal control system; MANUAL for manual elevator control, or HYDRAULIC for powered longitudinal control (flying tail). When this switch is placed in MANUAL, hydraulic pressure to all components of the powered longitudinal control system is shut off, and control is automatically shifted to manual operation.

EMERGENCY MANUAL CONTROL.

A push-pull knob (45, figure 1-3), labeled PUSH, is mounted on the inboard side of the stabilizer control selection panel on the left console. The emergency manual control knob provides a rapid means of selecting manual control in an emergency and also allows full electrical trim movement of the stabilizer. This knob normally remains in the up position, and must remain up for the powered longitudinal control system to be operable. The actual reversion to manual control is accomplished in the same manner as mentioned above; i.e., by shutting off hydraulic pressure to all the components of the powered longitudinal control system.

Note

The control stick should be near neutral when shifting from powered to manual control. If the control stick is held aft when reverting to manual control, the stabilizer cannot return to its hydraulic neutral position before pressure is shut off. This will result in a brief out-of-trim condition if the powered control system is again selected.

ABSOLUTE STABILIZER POSITION INDICATOR.

This indicator is mounted on the upper left side of the instrument panel. It is calibrated in degrees to indicate absolute stabilizer position. The range of the indicator is 0 to 4-1/2 degrees airplane nose-down, and 0 to 6-1/2 degrees airplane nose-up. On the manual control system, the indicator always denotes the trim position of the stabilizer and will move only when the stabilizer trim setting is changed. When operating on the powered longitudinal control system, the indicator continuously denotes the true stabilizer position and moves not only when the stabilizer trim adjustment is changed, but also whenever the control stick is displaced from neutral. Thus, when operating on the powered longitudinal control system, IT IS IMPORTANT TO REALIZE THAT THE ABSOLUTE STABILIZER POSITION INDICATOR DENOTES TRIM POSITION ONLY WHEN THE CONTROL STICK IS IN NEUTRAL AND NO FORE OR AFT CONTROL STICK FORCE IS BEING APPLIED.

ADJUSTABLE STABILIZER.

Longitudinal trim is controlled by switches through which the electric motor of the stabilizer actuator is energized. Trim is obtained through movement of the horizontal stabilizer by the electrically operated component of the actuator. The electric motor operates on d-c power from the essential bus. Limit switches in the stabilizer actuator restrict the stabilizer movement within the limits of its electrical adjustment range. The normal range is 2-1/2 degrees airplane nose-up and 1-3/4 degrees airplane nose-down with flaps up, and 6 degrees airplane nose-up and 1-3/4 degrees airplane nose-down with flaps down.

Note

The 6 degrees airplane nose-up and 1-3/4 degrees airplane nose-down limits are established by actual or absolute stabilizer position. Therefore, if the stabilizer is held away from these limits hydraulically while it is trimmed towards them electrically (pulling aft on the stick while trimming nose-down or vice versa), the limits can be exceeded and the trim can be run to the limit of actuator travel (6-1/2 degrees airplane nose-up or 4 degrees airplane nose-down).

AUTOMATIC RETRIM SYSTEM.

The purpose of automatic retrim is to eliminate excessive airplane nose-up trim change when the flaps are retracted. When the flaps are being retracted, with a trim setting of more than 2-1/2 degrees airplane nose-up, the stabilizer will automatically be repositioned to 2-1/2 degrees airplane nose-up trim. Once the flaps are up, the airplane nose-up trim is limited to 2-1/2 degrees. The operation of the automatic retrim system is dependent upon the trim and longitudinal control systems selection. Automatic retrim and the 2-1/2 degrees airplane nose-up limit is available only when the emergency manual control knob is in the normal (up) position, the electric trim switch is set to NORM., and the flaps have been retracted or are being retracted. The 2-1/2 degree limit becomes 6 degrees when the emergency manual control knob has been depressed or when the electric trim selection switch is set to EMERG.

ADJUSTABLE STABILIZER TRIM CONTROLS.

ELECTRICAL TRIM SELECTOR SWITCH.

The guarded electrical trim selector switch (42, figure 1-3) is mounted on the stabilizer control selection panel on the left console. It has three positions, NORM., OFF, and EMERG. This switch permits the selection of the normal or the emergency trim circuits. In the NORM. position, the stabilizer is adjusted by the momentary thumb switch on the control stick grip. In the OFF position, neither the normal nor emergency circuit can be operated. In the event of an uncalled for trim change (which may be caused by fusing of the electrical wiring running to the actuator) that cannot be corrected by use of either the normal or emergency trim switch, set the switch to OFF. This will stop any further actuator movement. In the EMERG. position, the stabilizer is adjusted by the emergency trim switch on the left console.

Note

Automatic retrimming and the 2-1/2 degree airplane nose-up limit are not available if the electrical trim selector switch is set to EMERG.

NORMAL TRIM SWITCH.

The momentary thumb switch (16, figure 1-4) on top of the control stick grip is normally used to trim the adjustable stabilizer. It is moved forward for airplane nose-down trim, aft for airplane nose-up trim, and is spring loaded to return to the central (off) position.

Note

Lateral movement of this switch controls the wing trimmer.

EMERGENCY TRIM SWITCH.

In the event of an open or short circuit in the normal control circuit, the electrical trim selection switch should be set to EMERG. and the stabilizer position adjusted by using the emergency trim switch. This momentary type toggle switch (41, figure 1-3) is mounted on the stabilizer control selection panel on the left console. It is moved forward for airplane nose-down trim, aft for airplane nose-up trim, and is spring loaded to return to the central (off) position.

RUDDER CONTROL SYSTEM.

The rudder pedals are linked to the rudder by cables and push rods. The position of the rudder is determined by conventional movement of the pedals and also by the electronic yaw damper system, if installed.

RUDDER TRIM TAB SYSTEM.[1]

An electrically operated rudder trim tab on the lower portion of the rudder is provided for directional trim. The limit of tab travel is 5 degrees to the left and 5 degrees to the right. Limit switches in the system automatically shut off electrical power when the tab reaches the limits of its travel. The tab is controlled by a rotary switch on the left console. A rudder tab position indicator mounted on the left console shows the position of the trim tab.

RUDDER TRIM TAB CONTROL SWITCH.

This five position rotary switch is marked NOSE LEFT and NOSE RIGHT and is mounted on the left console. The positions each side of neutral, as well as neutral, are off positions. Turning the switch to its extreme NOSE LEFT or NOSE RIGHT setting actuates the tab to obtain the desired directional trim.

RUDDER TRIM TAB POSITION INDICATOR.

This indicator is mounted on the left console. It is calibrated in degrees, with a range of 0 to 5 degrees nose left and 0 to 5 degrees nose right trim.

YAW DAMPER SYSTEM.[2]

The electronic yaw damper system eliminates lateral-directional oscillations which are sometimes encountered in swept wing airplanes. These oscillations are apparent to the pilot as forms of snaking, Dutch roll, or wallowing. The yaw damper senses small yawing motions and automatically displaces the rudder to correct the disturbances before they become apparent to the pilot.

The power requirements of the yaw damper system are 115 volt, 400 cycle, 3 phase ac for the amplifier and 28 volt dc for the servo unit motor. The ac is supplied by either inverter. The dc is supplied by the main bus and is controlled by the yaw damper power switch.

The yaw damper system consists essentially of two units: an amplifier and a rudder servo unit, and the control switches. The amplifier unit detects yawing velocity by means of a rate gyro and converts the mechanical displacement of the gyro into electrical signals. These signals are sent to the rudder servo unit, where the signals energize and also control the action of two clutches in the rudder servo unit. The rudder servo unit consists of the d-c motor, reduction geared to the two clutches which are geared to a common output shaft. This shaft drives a capstan connected to the rudder through cables. The rudder is deflected to correct the yaw disturbances detected by the rate gyro.

Note

The yaw damper indirectly activates the rudder-to-pedal cable system and causes the rudder pedals to move as the rudder is deflected by the yaw damper. This movement of the pedals may seem strange to the pilot on early familiarization flights, but will become virtually unnoticeable as the pilot gains experience in an airplane equipped with this system.

The yaw damper system is controlled by two pilot operated switches: the yaw damper power switch and the sensitivity selector switch. In addition, there are two automatic cut-out switch arrangements which disconnect the yaw damper whenever its action conflicts with pilot effort. The yaw damper normally operates whenever the yaw damper power switch is in the ON position, except when it is disconnected by the cut-out switches. One set of cut-out switches is associated with the rudder pedals and disconnects the yaw damper whenever the pilot applies a force greater than 45 pounds on either or both rudder pedals. This permits the pilot to yaw the airplane intentionally for control in cross-wind take-offs or landings.

Note

Pilots are cautioned not to rest their feet on the rudder pedals during straight flight so heavily that the 45 pound cut-out switches keep the yaw damper continuously disconnected.

The other set of cut-out switches is associated with lateral motion of the stick, and disconnects the yaw damper whenever the stick is displaced more than 1-1/2 inches from neutral. The yaw damper remains disconnected for two seconds after the stick has been returned to the 1-1/2 inch band on either side of neutral. This lateral cut-out feature allows a turn entry or roll reversal, permitting an intentional turn entry away from an established flight path. Since the yaw damper may be temporarily disconnected by lateral motion of the stick during a turn entry or roll reversal, appropriate rudder should be used to coordinate these maneuvers exactly as though no yaw damper was installed. The yaw damper sensitivity selector switch has two settings, LOW GAIN and HIGH GAIN, which allow the pilot to choose the sensitivity most comfortable to him under a given set of flight conditions. With the HIGH GAIN setting, the effectiveness of the yaw damper for oscillation elimination is good; however, in rough, turbulent air, particularly at low flight speeds, the motion of the rudder pedals may be annoyingly large. With the LOW GAIN setting, the effectiveness of the yaw damper is reduced, but in rough, turbulent air, particularly at low speed, the motion of the rudder pedals will be more tolerable.

YAW DAMPER POWER SWITCH.

This OFF-ON switch is mounted on the right cabin rail

[1] F9F-6 airplanes ser No. 128068 and subsequent and F9F-6P airplanes ser No. 127473 and subsequent.

[2] F9F-6 airplanes ser No. 130794 and subsequent.

just below the windshield frame. The yaw damper system is made operative by setting the yaw damper power switch to ON. This completes the circuit between the main bus and the rudder servo unit d-c motor.

YAW DAMPER SENSITIVITY SELECTOR SWITCH.

The amount of yaw correction provided by the system is determined by the setting of the yaw damper sensitivity selector switch located adjacent to the yaw damper power switch. The switch may be set to LOW GAIN or HIGH GAIN, depending upon the degree of correction desired by the pilot.

SURFACE CONTROLS LOCK.

The controls are locked by utilizing the pilot's harness and a cable assembly. The harness is used to secure the control stick and the cable assembly is attached to each rudder pedal and also secured to the control stick (figure 1-9).

FLAPS.

The wing flaps are hydraulically operated and are controlled by a lever (16, figure 1-3) on the left console. They can assume either of two positions, up or down. The down position is used for both take-off and landing. Flap surfaces are on the trailing edge of each wing and beneath the fuselage. The wing flaps are referred to as ouboard flaps and the fuselage flaps as the inboard flaps, using the landing gear as datum points. In operation, when the flap control lever is actuated, it directs hydraulic pressure to the flaps actuating cylinders, which, in turn, move both the outboard and inboard flaps (30 degree outboard flaps and 40 degree inboard flaps full deflection). The flaps position is indicated on the wheels and flaps position indicator on the main instrument panel (2, figure 1-3A).

Note

To actuate the flaps, first move the control lever inboard to free it from its detent before attempting to select the DOWN position.

When flap control lever is set at DOWN, wing slats extend simultaneously with the flaps.

SPEED BRAKES.

The hydraulically operated speed brakes are installed beneath the fuselage and on the trailing edges of the inboard flaps and have a limit extension of 75 degrees and 40 degrees, respectively. The speed brakes may be extended to any intermediate position but cannot be extended or retracted separately.

SPEED BRAKES NORMAL CONTROL.

The speed brakes are controlled by a momentary type switch on the inboard side of the throttle quadrant, spring loaded to OFF (20, figure 1-2A). Moving the switch to UP or DOWN energizes a solenoid valve which directs hydraulic pressure to the speed brake cylinders. Hydraulic pressure will keep moving the speed brakes as long

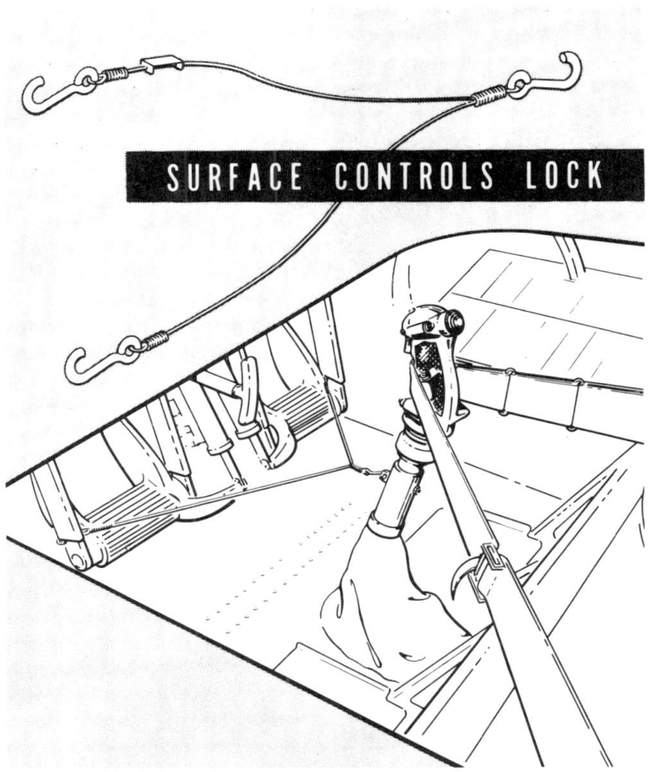

Figure 1-9. Surface Controls Lock

as the switch is held in the selected position (UP or DOWN) and will stop when the switch is released (to OFF) or when they reach the limit of their travel.

Note

Normally, the landing flaps and speed brakes cannot be down simultaneously. If the flaps are extended while the speed brakes are down, the latter will retract automatically. However, for a wheels-up emergency landing, the speed brakes can be extended by the normal control when the landing flaps are extended by means of the speed brakes emergency landing override switch. (See Speed Brakes Emergency Landing Override Switch paragraph.)

SPEED BRAKES EMERGENCY CONTROLS.

SPEED BRAKES MANUAL OVERRIDE CONTROL.

If normal operation fails, the speed brakes manual override control (3, figure 1-2A), a "T" handle on the left side of the after cabin bulkhead, may be used to retract the speed brakes. Pulling this handle will mechanically operate the solenoid valve which directs hydraulic pressure to raise the speed brakes.

Note

A special speed brakes reset type circuit breaker is mounted on the left console (22, figure 1-3C) outboard of the normal control. If the speed brake circuit breaker will not stay set, check that speed brake switch is in the OFF position. If speed brakes are out, retract by use of the speed brakes manual override control.

SPEED BRAKES EMERGENCY LANDING OVERRIDE SWITCH.

This two position, guarded toggle switch, mounted on the aft portion of the left console (27, figure 1-2A), is safety-wired in the normal position. In the normal position, the speed brakes will be raised automatically whenever the landing flaps are lowered. Breaking the safety wire, raising the guard and setting the switch to the emergency position, with landing flaps down, permits the speed brakes to be lowered using the normal speed brakes control switch. This procedure permits lowering both the flaps and speed brakes in an emergency such as a wheels-up landing.

SPEED BRAKES POSITION INDICATOR.

A speed brakes position indicator is installed on the left side of the instrument panel (8, figure 1-3A). When the speed brakes are fully retracted, the word UP appears on the indicator. When the speed brakes are in any intermediate position or full down, the word DOWN appears on the indicator. A barber pole flag appears on the dial face when power to the indicator is cut off.

SLATS.

Slats are installed on the leading edge of the wing to augment the slow speed controllability of the aircraft. They are hydraulically operated and extend (forward of the leading edge) simultaneously with the flaps when the latter are lowered. The slats retract when the flaps are raised.

LANDING GEAR.

NORMAL CONTROL.

The landing gear control, a two position lever operating a hydraulic selector valve, is on the instrument panel bulkhead forward of the left console (44, figure 1-3A). When the lever is moved to DOWN, after releasing the "T" latch, hydraulic cylinders open the wheel doors, lower and lock down the main and nose wheels and then close the main wheel doors again. The doors are operated by separate cylinders. When the lever is moved to UP, action is reversed and special locking cylinders lock the main wheels in the up position. A solenoid operated latch is installed to prevent inadvertent wheel retraction when the airplane is on the deck. When airborne, the struts extend, releasing the latch, which permits moving the control lever to UP. Extension or retraction of the landing gear may be accomplished by the normal control, using the auxiliary hydraulic pump in the event of normal hydraulic pump failure.

Note

When the landing gear control handle is moved to DOWN, the guns are safetied.

EMERGENCY CONTROLS.

EMERGENCY EXTENSION CONTROL HANDLE.

A red "T" handle, just outboard of the normal control (1, figure 1-3A), when pulled out, actuates a cable to open a compressed air bottle and direct air pressure to the cylinders to open the doors and lower and lock the wheels (and again close the main wheel doors). The handle shaft is notched to provide a lock when it is pulled out to release position. IT MUST BE LOCKED WHEN PULLED TO THE RELEASE POSITION.

WARNING

Effective on airplanes with F9F Aircraft Service Change 217 incorporated, pulling the emergency landing gear "T" handle will cause the normal hydraulic system to be dumped to the return line. Therefore, in a situation which will require landing gear emergency extension, efforts should be made to lower the flaps prior to extension of the landing gear.

Note

Once the emergency control has been used, the gear cannot be retracted until normal operation has been restored by a deck crew by reservicing the system.

DOWN LOCK SOLENOID RELEASE.

A thumb operated release knob is located just forward of the instrument panel bulkhead on the left side, just inboard of the normal control (42, figure 1-3A). If electrical failure should make the down lock solenoid inoperative, pushing the knob outboard will clear the control lever to raise the gear for a wheels-up landing.

INDICATORS.

WHEELS AND FLAPS POSITION INDICATOR.

A standard indicator which shows wheels down and locked or wheels up and locked is on the left side of the main instrument panel (2, figure 1-3A).

LANDING GEAR UNLOCKED WARNING LIGHT.

A red light in the translucent handle of the landing gear control lever (44, figure 1-3A) glows when the gear is not locked in the up or down position.

BRAKES.

NORMAL CONTROLS.

The brake pedals operate power assist brake valves (mounted on the pedals) to control hydraulic pressure to the disc type brakes. A special hydraulic reservoir (connected to the main system) is installed for the brakes so that brake operation is available in the event of hydraulic system failure.

EMERGENCY CONTROL.

The emergency brake red "T" handle is on the left console, aft (26, figure 1-2A). The handle is pulled up to release and inboard to lock (the handle shaft is notched to provide a lock for the release and lock position). When pulled, the handle actuates a cable to open a compressed air bottle which acts on the brake cylinders.

Note

Once the emergency "T" handle is used, the brakes are set full on. Release and normal operation must be effected by a deck crew.

ARRESTING HOOK.

NORMAL CONTROLS.

HOOK DOWN AND BARRIER GUARD CONTROL.

The hook control, on the instrument panel bulkhead forward of the right console (26, figure 1-3A), operates a chain and cable assembly to the hook and a cable assembly to the barrier guard. The hook rides on a carriage and roller assembly and the control must be pulled and released repeatedly (three or four times) until the hook is fully extended and locked down. The barrier guard is a spring loaded square tube which, when the control handle is pulled, projects up from the fuselage just forward of the windshield. A switch on the release lowers the tail skid when the hook is lowered. The hook and barrier guard must be restowed by the deck crew.

CAUTION

When test operating the system, ensure that personnel are clear of the hook and barrier guard.

HOOK UP CONTROL.

This push button switch (27, figure 1-3A) is on the instrument panel bulkhead forward of the right console, adjacent to the hook down and barrier guard control. When the button is pushed, a solenoid operated hydraulic valve is actuated to direct pressure to a single action cylinder which raises the hook clear of the deck to permit taxiing forward of the barrier and arresting cables. (This cylinder action retracts the tail skid simultaneously through a cable linkage.) This solenoid will not be energized if the airplane is airborne; a switch on the right main landing gear nutcracker completes the circuit only when the airplane is on the deck. When the fuel master switch is set to OFF to shut down the engine, the solenoid valve releases pressure and the hook drops to the deck; this action serves as a signal to the deck crew that the engine has stopped and they may push the hook forward into the stowed position (after unlocking the hook carriage by pulling out the recessed handle on the left side of the fuselage tail).

INDICATORS.

HOOK POSITION WARNING LIGHT.

A red press-to-test warning light on the main instrument panel bulkhead forward of the right console (22, figure 1-3A) glows when the hook is in any intermediate position between full up and full down. The light goes out when the hook is either fully extended or retracted.

APPROACH LIGHT.

When the wheels are down and locked and the hook is lowered, the approach light is lighted (steady); if the wheels are down and locked and the hook is *not* lowered the approach light flashes.

TAIL SKID.

The skid is interconnected mechanically with the arresting hook. The hook raising cylinder raises the skid through a cam and cable arrangement and the skid is lowered when pressure to the cylinder is released. Pressure to the cylinder is controlled by a solenoid valve operated by switches as follows.

a. A switch on the landing gear control causes the cylinder to extend and raise the skid when the gear is raised and releases pressure from the cylinder to cause the skid to lower when the gear is lowered.

b. A limit switch actuated by the throttle is utilized to raise the skid when the throttle is moved full forward to OPEN and to lower the skid when the throttle is retarded to 90 per cent power when the landing gear control is at DOWN. This control is installed for field carrier landing practice.

c. A switch on the arresting hook release overrides both the landing gear control and the throttle switches to lower the skid when the hook is lowered.

d. The hook raising switch button, when pressed to raise hook from deck after landing, also raises skid.

TAIL SKID CONTROL SWITCH.

A tail skid control switch equipped with a guard (23, figure 1-3A) is provided. When set to DOWN, after clearing the guard, the throttle actuated skid raising switch is overriden and the skid remains down until the landing gear is raised.

WING FOLDING.

NORMAL CONTROLS.

The folding and locking control assembly (figure 1-10) is on the aft end of the right console. The folding lever lies flush with the console and the crank is in its aft position when the wings are spread and locked. To fold the wings, swing the crank forward, push down on the plate marked PUSH, and raise the lever marked LIFT. Rotating the crank forward disengages the safety locks in the wing lock cylinders. A detent on the crank shaft then clears the lever so it can be lifted up to unlock the wing lock cylinders and to fold the wings. The lever is pushed down into the console to spread and lock the wings and the crank is rotated aft to engage the safety locks on the lock cylinders.

WARNING

Do NOT force the crank into the locked position if the normal amount of force does not cause the crank to swing aft. This may be an indication that the wing fold hinge lock pin is not in the locked position.

The lever operates a hydraulic selector valve to actuate folding and locking cylinders (one each—left and right). When pushed down (spread), the timer check valves open after the folding cylinders are fully retracted to permit pressure to actuate the lock cylinders. In the folding action (crank forward, lever lifted), the timer check valves permit pressure to flow to the folding cylinders after the locking cylinder pistons have been withdrawn.

CAUTION

Make certain the wing lock lock control crank is moved through its limit of travel before operating the wing folding control lever.

Never leave the wing folding control lever in an intermediate position between its wing folded and wing spread positions.

The wing folding or spreading cycle must be completed before reversing the direction of wing movement.

After wings are folded, it is recommended that wing fold jury struts be inserted before the engine is shut down.

Before spreading wings, ensure that dump valves are closed.

The crank of the assembly, when rotated aft, actuates a cable and bellcrank linkage which rotates notched pins to block the locking cylinder pistons in the locked position and pulls the red signal vanes out of sight when the pins are in place. When the crank is swung forward, the pins are rotated to clear the lock cylinder pistons for withdrawal (unlock) and the signal vanes appear.

INDICATORS.

Red signal vanes are installed on the upper surface of the wing stub leading edges just inboard of the folding axes. When the wings are spread and locked, the vanes are drawn out of sight of the pilot and into the wings by mechanical action, through cables and bellcranks from the locking control. When the wings are unlocked, the vanes protrude from the wings and are visible to the pilot.

INSTRUMENTS.

The majority of the instruments are installed on the main instrument panel (figure 1-3A). They can be classified into three groups; flight instruments, engine instruments and miscellaneous instruments. In the flight instrument group the altimeter rate of climb indicator, and airspeed indicator operate from the pitot static system; the turn and bank indicator, from engine compressor bleed air. The G-2 remote compass and gyro horizon indicator are powered by ac supplied by the inverter system. The radio compass is d-c powered from the main bus. In the engine instrument group, the tailpipe temperature indicator is d-c powered from the main bus. The fuel quantity indicator is powered by dc from the main bus and ac from the inverter system. The fuel flowmeter and oil pressure indicator are powered by ac from the inverter system. The tachometer indicator is powered by ac from a self-contained source (tachometer generator). The miscellaneous group of instruments comprises the wheels and flaps position indicator, the speed brakes position indicator, the stabilizer position indicator, and the rudder trim tab position indicator, which are powered by dc from the essential bus. Included in this group are the accelerometer, which is internally mechanical, and the Bourdon tube type hydraulic pressure gages.

G-2 REMOTE COMPASS.

The G-2 compass combines the advantages of the remote indicating compass and the gyro compass into one instrument, providing a compass-slaved directional gyro system compass reading, which corrects automatically for gyro precession, and eliminates compass oscillation and northerly turning error. A correspondence dial, located in the center of the master compass indicator, gives an unstabilized remote compass reading, and the outer dial gives a stabilized gyro compass heading, free from drift. A control switch, located on the electrical control panel on the right console (39, figure 1-3B), provides compass slaving of the instrument or permits its use as a free directional gyro. (See Navigation Equipment paragraph, Section IV, for operation.)

MAXIMUM ALLOWABLE AIRSPEED INDICATOR.

The maximum allowable airspeed indicator is located on the upper row of instruments to the left of the center line and has a range of 0 to 650 knots in increments of ten (40, figure 1-3A). It is connected to both static and pressure lines. The indicator is a combination instrument consisting of two entirely different mechanisms, the pointers of which indicate on the same dial. One of these instruments is a conventional airspeed indicator and the other, usually called a Mach mechanism, is an altimeter calibrated to show maximum allowable airspeed for a given Mach number. The Mach mechanism is preset to the highest flight Mach number for the airplane. This number appears in the small window of the dial. The indicated airspeed corresponding to the safe Mach number is then indicated by a warning pointer which is wider than the airspeed pointer and is painted with red and yellow stripes. As the airplane ascends, the warning pointer indicates lower and lower airspeeds, because the indicated airspeed for a given Mach number diminishes with altitude, since the value of the speed of sound decreases with the increases in altitude. The maximum airspeed pointer is used by the pilot to determine the airspeed corresponding to the limiting Mach number of the airplane in unaccelerated flight. As long as the pointers do not cross, he is flying safely. Adjustment screws on the indicator are used to set stops for the maximum allowable airspeed and critical Mach number. An airspeed correction card holder and an airspeed correction card are located on the right windshield frame.

AIRSPEED AND MACH NUMBER INDICATOR.

Note

The airspeed and Mach number indicator will, when available, replace the maximum allowable airspeed indicator.

The airspeed and Mach number indicator is located on the instrument panel in the upper row of instruments to the left of the center line. The indicator is a combination instrument, connected to both static and pitot pressure lines, consisting of an airspeed mechanism and an altitude mechanism. The indicated airspeed mechanism drives a pointer to indicate airspeed on a fixed dial. The pointer also indicates the Mach number on a movable Mach number scale which is driven by the altitude mechanism.

The gearing between the moving scale and the altitude mechanism is such that the Mach number will be indicated by the pointer on the moving scale at any combination of indicated airspeed and altitude within the range of the instrument. The airspeed indicator has a range of 80 to 650 knots and the Mach number scale has a range of 0.50 to 2.0. The indicator has two triangular indexes, a Mach number setting index and an airspeed setting index. The setting of the Mach number index and the airspeed index is made by the adjustment knob located in the lower left corner of the bezel. The Mach number index is set by depressing and turning the knob and setting the index to the desired Mach number. The Mach number index is not necessarily set at the limit Mach number of the airplane. It is for the pilot's convenience in flying any constant Mach number within the airplane limitations. The airspeed indicator index is set by turning the knob, without depressing it, and setting the index to any desired airspeed within its range of 80 to 145 knots.

Note

Either index can be set without changing the setting of the other index.

The following two conditions are given as examples of the operation of the indicator. As the airplane ascends at a constant airspeed, the pointer will remain stationary and the Mach number scale will rotate, indicating an increasing Mach number. As the airplane ascends at a constant Mach number, the Mach number scale and the pointer rotate, with the pointer maintaining its position on the Mach number scale, indicating a constant Mach number and decreasing airspeed. This is due to the fact that the speed of sound decreases with increasing altitude.

FAST ERECTING GYRO HORIZON.[1]

A fast erecting gyro horizon indicator is located on the upper row of the main instrument panel to the right of the center line (31, figure 1-3A). A warning light and control switch are located outboard on the electrical control panel on the right console (40, 42, figure 1-3B). The power unit is installed in the fuselage. This installation provides an immediate indication of the airplane's attitude.

OPERATION.

Normal operation of the installation is automatic upon setting the battery switch to BAT. & GEN. When power is turned on, the indicator gyro is energized from a transformer and the electromagnet is energized from a rectifier at the same time. This combined action erects the gyro and brings the motor up to speed in less than one minute. At the end of this time the electromagnet is deenergized and the gyro rotor circuit is switched to the normal operating voltage. When the power unit is first energized, the red warning light glows and will go out when the magnet is deenergized. If the light glows for longer than one minute, indicating power unit malfunction, the switch must be set to OFF; this is done to prevent the electromagnet from remaining energized unnecessarily.

GYRO HORIZON INDICATOR.[2]

A gyro horizon indicator is located on the upper row of the main instrument panel to the right of the center line (31, figure 1-3A). This instrument provides an immediate indication of the airplane's attitude when properly erected.

The indicator has a pitch scale visible through a rectangular window located in the center of the instrument.

[1] F9F-6 airplanes ser No. 126257 through 128130 and F9F-6P airplanes ser No. 127473 through 131255.

[2] F9F-6 airplanes ser No. 128131 and subsequent and F9F-6P airplanes ser No. 134446 and subsequent.

The pitch scale has a range of 0 to 80 degrees for indicating both positive and negative airplane pitch attitudes.

OPERATION.

The gyro horizon indicator should be caged immediately after the battery switch is set to BAT. & GEN. The PULL TO CAGE knob should be drawn away from the face of the instrument smoothly. (It should never be pulled hard.) It will stop momentarily when the bank caging is first engaged. Continue pulling until the horizontal and banking index bars indicate 0 pitch and 0 roll, ± 3 degrees. The time required to hold the PULL TO CAGE knob out before the index bars stabilize depends upon the position of the gyro; however, the longest time will be approximately 10 seconds. As soon as the gyro is caged and the knob reaches the limit of its travel, it should be released QUICKLY. By means of the small knob on the left side of the bezel, the miniature airplane may be raised or lowered for any desired fore-and-aft attitude of the airplane. When the battery switch is set to OFF, or if there is a partial or complete power failure, an OFF flag will appear on the dial face.

CAUTION

Do not cage the gyro horizon indicator in flight unless the airplane is flying straight and level, or the gyro will give an incorrect indication.

In some instances, after prolonged turns or aerobatics, the gyro horizon may have slight errors due to the accumulated effect of centrifugal forces sensed as gravitational forces by the instrument. The gyro will immediately begin to correct these errors, once true gravitational forces are sensed. The pilot may desire to speed erection by caging the gyro manually—but this must not be done unless such error is known to be excessive. The pilot must realize two things: (1) He must realize that his sense of vertical after turns or aerobatics is less accurate than the indication of the instrument. (2) He must realize that the instrument cages to the ATTITUDE of the aircraft, NOT to the TRUE VERTICAL.

If, under normal conditions, the gyro horizon should give readings which appear inaccurate or disagree with other instruments (such as the turn and bank indicator or the rate of climb indicator), the pilot should determine whether or not the instrument is malfunctioning. He should check the OFF flag for indication of improper operating power and take corrective action if necessary. If the gyro horizon is still inaccurate, it should be removed and replaced. Excessive caging will not correct attitude horizon indicators which are malfunctioning; excessive caging will cause wear, unnecessary strain, and possibly jeopardize the safety of the pilot by rendering the instrument inoperative.

STAND-BY COMPASS.

The stand-by compass (13, figure 1-3A) is mounted above the instrument panel to the right of the sight unit. Two correction cards are mounted below the instrument panel to the right of the center line. One card shows compass calibration for normal operating condition with all equipment in operation; the second card gives the calibration with the battery and generator switch in the OFF position.

CANOPY.

NORMAL CONTROLS.

CANOPY CONTROL—NORMAL HYDRAULIC.

This handle is on the left side of the cabin, just forward of the canopy track. The handle is moved forward to CLOSE and aft to OPEN. The control handle operates a selector valve to actuate a hydraulic cylinder on the fuselage deck to move the canopy forward and aft (3, figure 1-3A).

CAUTION

Do not return canopy control handle to neutral when canopy is moved to either full open or full closed.

Keep head and hands inside canopy during operation. If canopy does not close fully, it indicates faulty latching. Do not open in flight above limiting airspeed except in an emergency, as it may be lost.

CANOPY CONTROL—GROUND OPERATION.

This lever is on the left side of the fuselage aft of the line of the seat armor plate, just below the edge of the canopy. The control is operated by pressing in the forward end with the thumb and then rotating the handle up to OPEN or down to CLOSED. The lever actuates a switch to start the auxiliary hydraulic pump and operates the selector valve to direct pressure to the canopy cylinder.

WARNING

See that personnel are clear of flaps and dive brakes when operating canopy ground operation control.

EMERGENCY CONTROL.

The normal canopy control described above incorporates an emergency canopy control lever labeled emergency open. Folding this handle into a slot in the normal control handle permits the entire control assembly to be moved beyond OPEN and into the EMERG OPEN position. This releases air pressure from an air bottle system into the canopy control hydraulic cylinder to open the canopy (4, figure 1-3A).

CANOPY GROUND RELEASE.

The canopy may be removed by using the handgrip on the aft end. Pressing down the handgrip unlatches the canopy from the actuating cylinder.

PILOT'S EJECTION SEAT.

NORMAL CONTROLS.

SEAT HEIGHT CONTROL.[1]

This lever is mounted on the right console above the main circuit breaker panel. It is moved aft to DOWN to lower the seat and forward to UP to raise it. The lever operates a selector valve controlling pressure to the height adjusting hydraulic cylinders behind the seat. The control is spring loaded and when released, locks the seat at the selected position (26, figure 1-3B).

SEAT HEIGHT CONTROL.[2]

Seat height is adjusted by an electrical actuator controlled by a three position momentary switch, with a center (off) position, mounted on the right side of the seat. The two momentary positions are DOWN and UP (1, figure 1-4).

HARNESS LOCK.

The harness inertia reel lock is mounted on the forward left side of the seat. The lever is moved AFT TO SLACK OFF the harness in order to lean forward, etc; the lever is moved FORWARD TO LOCK the harness (13, figure 1-4).

HEADREST.

The headrest is on the ejection seat assembly. On certain airplanes[3] it may be swung forward for catapulting by releasing the latch on the right side, pulling the headrest forward and resetting the latch. On later airplanes[4], the ejection seat has been redesigned to incorporate a fixed headrest.

EMERGENCY CONTROLS.

PRE-EJECTION LEVER.

This lever is on the left side of the cabin below the canopy track (15, figure 1-2A). It is PULLED INBOARD AND PUSHED DOWN to operate and then PUSHED OUTBOARD TO LOCK. When the lever is operated, the canopy is jettisoned (by pressure from an air bottle), the seat is lowered (by pressure from the brake air bottle), the knee braces on the seat sides are moved up into position, and the firing charge safety is pulled.

WARNING

The lever MUST BE PUSHED OUTBOARD TO LOCK DOWN.

EMERGENCY EJECTION SEAT ARMING CONTROL.

On airplanes equipped with fixed headrest seats, an emergency ejection seat arming control is provided. Under certain extreme flight conditions, such as high negative "g", it may be difficult to reach and actuate the pre-ejection lever. For this condition, or one where the canopy fails to jettison, using the emergency ejection seat arming control will permit ejection through the canopy. This control is a red handle on the left side of the headrest and is attached to the cable which normally pulls the safety pin in the seat catapult firing mechanism. Pulling the control forward past the pilot's left ear pulls the safety pin and arms the seat. It does not lower the seat nor does it release the knee braces. For this type of ejection, hold knees together to prevent injury.

WARNING

Ejection through the canopy is feasible only with the canopy in the fully closed position. The canopy cannot be closed if the pre-ejection lever has been actuated.

SEAT CATAPULT FIRING CONTROL (FACE COVER HANDLE).

This handle (5, figure 1-4) is just above the pilot's headrest and is attached to the face cover which is stowed on a roller behind the headrest. After operating the pre-ejection lever, the pilot slides his feet aft onto the stirrups, sets his knees against the braces and pulls the

[1] F9F-6 airplanes ser No. 126257 through 128239 and F9F-6P airplanes ser No. 127473 through 128306.
[2] F9F-6 airplanes ser No. 128240 and subsequent and F9F-6P airplanes ser No. 128307 and subsequent.
[3] F9F-6 airplanes ser No. 126257 through 127422 and F9F-6P airplanes ser No. 127473 through 127480.
[4] F9F-6 airplanes ser No. 127423 and subsequent and F9F-6P airplanes ser No. 127481 and subsequent.

handle forward and down to bring the face curtain over his face. This action locks the inertia reel and fires the charge to eject the seat from the airplane. The seat is ejected from the airplane by the catapult unit which consists of two mated tubes and an explosive charge. The inside tube is attached to the cabin floor. The outside tube, which carries the charge, is attached to the seat back.

AUXILIARY EQUIPMENT.

The following systems are discussed in Section IV:

Cabin Air Conditioning System
Communication and Associated Electronic Equipment
Lighting Equipment
Oxygen System
Navigation Equipment
Armament
Armament Control System
Gun Camera
"G" Suit Equipment
Photographic Equipment (F9F-6P)
Automatic Pilot (F9F-6P)

NAVY

SECTION II

TABLE OF CONTENTS

SECTION II NORMAL PROCEDURES

BEFORE ENTERING THE AIRPLANE.

a. See Section V and service directives and technical orders for limitations imposed on this airplane.

b. PLAN YOUR FLIGHT. Check aerology and applicable NOTAMS and other directives. Use the operating data contained in Appendix I to determine airspeeds, distances, fuel reserve, etc.

c. Determine weight and balance status of airplane:—Obtain take-off and anticipated landing gross weights. Have useful load checked. Be sure proper ammunition load or ballast is aboard.

Note

Refer to the Handbook of Weight and Balance, AN 01-1B-40, for loading information.

EXTERIOR INSPECTION.

Consult Yellow Sheet for status of airplane and make sure airplane has been properly serviced. For complete exterior inspection, see figure 2-1.

ENTRANCE.

For instructions on entering the airplane, see figure 2-2.

CAUTION

The boarding ladder cannot be stowed from the cabin. Before flight, ensure that it is properly stowed by the ground crew.

CANOPY AND EJECTION SEAT CHECK.

Before becoming seated in the airplane, check canopy for normal position (not jettison position).

a. Check for proper attachment of the firing cable from the face curtain to the firing yoke.

b. Determine that the safety pin is in place and that the firing mechanism is cocked and ready for firing.

ON ENTERING AIRPLANE.

INTERIOR CHECKS—ALL FLIGHTS.

a. All switches off or safe (gun switches).

b. Remove controls lock (bridle) and stow.

c. Control stick and pedals for free movement and for full throw.

d. Oxygen ON (oxygen supply control at regulator). Check equipment.

e. External power sources connected (check with ground crew).

Note

There are three external power receptacles; engine starting, general equipment and electronic testing.

f. Clock set.

g. Altimeter set.

h. Cabin clear, everything stowed; seat belt and

EXTERIOR INSPECTION

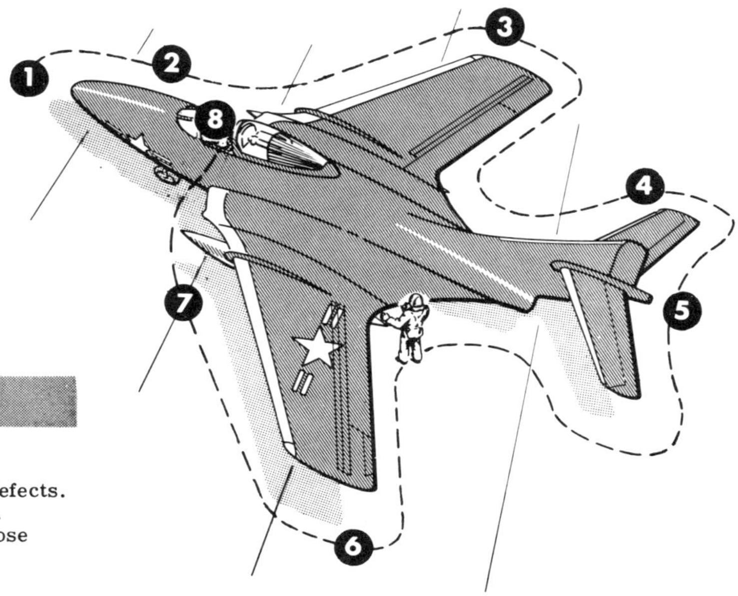

1 NOSE

a. Overall view of airplane.
b. Landing gear chocked.
c. Overall view of nose gear indicates no obvious defects.
d. Nose gear tire and shock strut properly inflated.
e. Check nose for positive lock and air bottles in nose wheel well for proper pressure.
f. Left and right gun barrels unplugged.
g. Camera window covers off and windows clean (F9F-6P airplanes only).

2 FUSELAGE - RIGHT SIDE

a. Access doors closed.
b. Intake duct unplugged.
c. Overall view of main landing gear indicates no obvious defects.
d. Main gear tire and shock strut properly inflated.
e. Speed brakes full up.
f. Inboard flaps and speed brakes hinge points and contour flush with fuselage and wing stub.

3 RIGHT WING

a. Overall view of wing upper and lower surfaces indicated no washboarding, wrinkles or cracks.
b. Wing fuel tank filler cap and door secure.
c. External stores secure. Filler cap on droppable fuel tank secure.
d. Navigation lights undamaged.
e. Slat completely retracted-maximum gap 1/16 inch.
f. Flaperon full flush.
g. Flaperon hinge points secured and safetied.
h. Flap full up (no trail). Hinge points secured and safetied.
i. Wing fairings in place.
j. Wing tank dump port unplugged.

4 STABILIZER

a. Overall view indicates no wrinkle or cracking.
b. Elevators move freely. Hinge points secured and safetied.

NOTE: On airplanes equipped with the flying tail, elevators will move as a unit. On airplanes equipped with the spring tab elevator system, the elevators are not linked together and each elevator must be checked seperately for freedom of movement.

c. Elevator tab hinge points secured and safetied.
d. Elevator tab play - no noticeable play permissible.
e. Stabilizer fairings intact, screws in place and no deformation of fairing itself.

5 TAIL SECTION

a. Overall view of fin and rudder indicate no wrinkle or crack.
b. Rudder moves freely. Hinge points secured and safetied.
c. Rudder trim tab (if installed) - no noticeable play permissible.
d. Navigation lights undamaged
e. Look up tailpipe to check condition of turbine blades and check shroud to see that there is no evidence of turbine blades rubbing.
f. Auxiliary air doors uncovered.

6 LEFT WING

a. Reverse procedure given for right wing.
b. Wing trimmer hinge points secured and safetied with cotter pin.
c. Play in trimmer not to exceed 1/8 inch.
d. Flaperon full flush with wing surface.
e. Wing fuel tank filler cap and cover in place.

7 FUSELAGE - LEFT SIDE

a. Inboard flaps and speed brakes hinge points secured and safetied and contour flush with fuselage and wing stub.
b. Speed brakes full up.
c. Overall view of main landing gear indicates no obvious defects.
d. Main gear tire and shock strut properly inflated.
e. Intake duct unplugged.
f. Access doors closed.
g. Fuselage fuel tanks filler caps and doors secure.

8 CABIN

a. Ejection seat for proper attachment of catapult firing and safety pin removal lanyards.
b. All switches off.
c. Controls unlocked.
d. Control bridle stowed.

Figure 2-1. Exterior Inspection Diagram

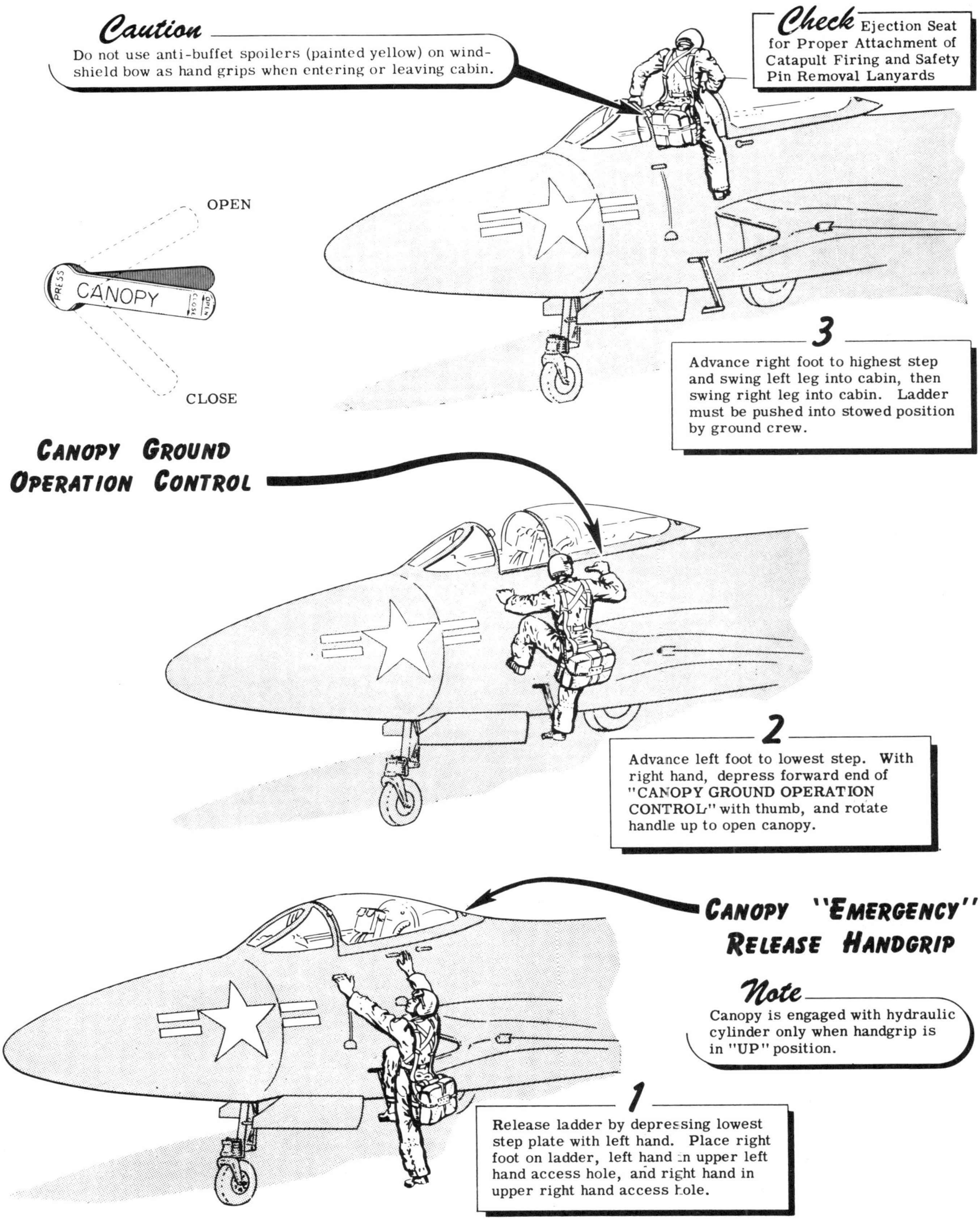

Figure 2-2. Entrance to Airplane

shoulder harness fastened; oxygen, "G" suit, and radio leads attached.

i. "G" suit valve HI or LO, as desired.

j. All circuit breakers in.

k. Battery switch—BAT. & GEN. Check battery voltage (21 volts minimum).

l. Check landing gear down and locked.

m. Pitot heater switch—ON (if necessary).

n.. Fuselage sliding nose locked (warning light out).

o. Check all press-to-test lights.

p. Fast erecting gyro horizon warning light. If light still glows after battery has been on one minute, set switch to OFF.

Note

Warning light cover may be removed for daytime operation.

q. G-2 compass control set to COMPASS CONTROL.

r. Fire warning lights (test lights and system).

s. Check operation of auxiliary hydraulic pump—1700 psi normal.

CAUTION

Do not operate pump any longer than necessary.

t. Emergency electric hydraulic pump for lateral control (for airplanes so equipped). Set flaperette system switch to EMER. ON & TEST. Move control stick to full travel right and left and check operation of flaperettes. Return switch to AUTO position after check.

Note

This check must be performed before starting engine and after wings are spread. There is no check for the engine driven flaperette pump until the engine is running.

u. Set stabilizer electric power switch to EMER. Trim nose-up and nose-down while checking indicator and visually check movement of the stabilizer in rear view mirror. IMPORTANT—RESET STABILIZER POWER SWITCH TO NORMAL AND REPEAT CHECK.

INTERIOR CHECK—NIGHT FLIGHTS.

Note

A dependable flashlight should be taken on all night flights in case of interior lights failure.

a. Operate instrument panel lights and console lights to BRIGHT.

b. Set exterior lights control switches to DIM and BRIGHT and operate master exterior lights selector switch. Set code switch to MAN and operate key; set code switch to AUTO and operate fuselage lights coding switch. Check indicator light.

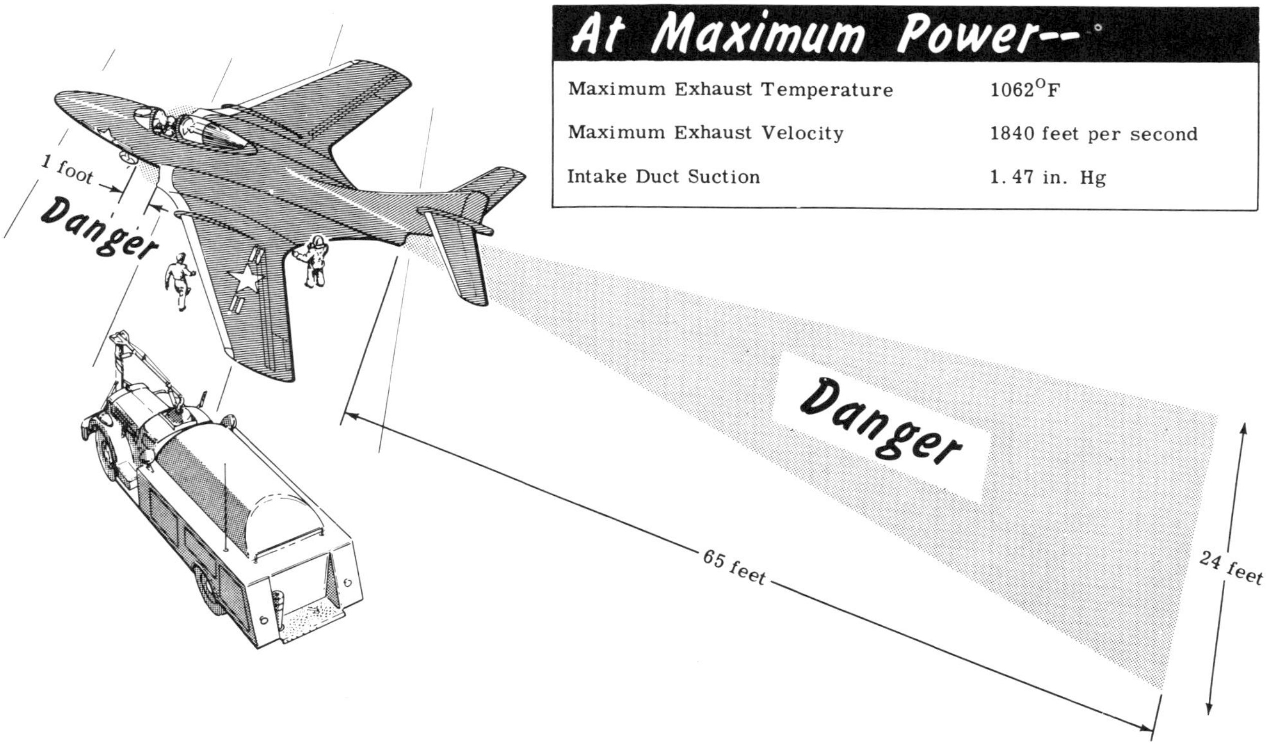

At Maximum Power--

Maximum Exhaust Temperature	1062°F
Maximum Exhaust Velocity	1840 feet per second
Intake Duct Suction	1.47 in. Hg

Figure 2-3. Danger Areas

c. Dim all warning lights as desired.

BEFORE STARTING ENGINE.

CAUTION

Check that airplane is turned into the wind, that personnel are clear of the tailpipe and intakes (see figure 2-3), and that external power is plugged in.

a. Wing trimmer at zero.

b. IMPORTANT—STABILIZER 3.5° airplane nose-up trim on flying tail airplanes or 5.5° nose-up on gear tab elevator airplanes.

c. Rudder trimmer at zero.

d. Set oxygen regulator as desired.

e. Hydraulic pressure control lever—ON.

f. Fuel quantity. Check push-to-test and low fuel level warning light.

g. Check that indicating pointer on tailpipe temperature indicator is set at 800°C.

h. Check that flaperette emergency system air pressure gage indicates 1900-2000 psi.

i. All circuit breakers, except inverter and gun charging, set (reset buttons in).

STARTING ENGINE.

CAUTION

Use only the following procedure to start the engine. Any divergence might result in a hot start, a condition which is seriously detrimental to the engine and which will abnormally shorten its life. See Hot Starts paragraph for precautions to be observed when a hot start is encountered.

Except in an emergency:

Do not exceed electric starter duty cycle of 30 seconds.

Do not accomplish more than three consecutive cycles (including clear engine) without allowing adequate cooling of the starter.

Note

Tachometer indicator shows per cent of rated rpm—100% = 11000 rpm.

STARTING PROCEDURE—J48-P-6A ENGINE.

a. Check that throttle is at CLOSED.

b. Battery switch at BAT. & GEN.

c. Set fuel master switch to ON—after a few seconds, check that fuel boost pressure warning light goes out, indicating that sufficient fuel pressure (6 psi minimum) has been built up.

Note

This is not a positive check of the boost pump. A malfunctioning pump may build up 6 psi fuel pressure with a closed fuel system and still not provide sufficient pressure under flow conditions.

d. Set engine start master switch to ON.

e. Set engine fuel system selector switch to START & EMER. Amber warning light will glow.

f. Set cranking switch to START and release.

g. When tachometer indicates 5-6% rpm, move throttle outboard to START, hold for three to five seconds, then move into its IDLE detent. (Do not move throttle beyond the IDLE detent during the start cycle). Light-up should occur and be indicated by a rapid rise in tailpipe temperature to approximately 650°C. The temperature limit for idling is 565°C. After starting, allow engine to run at idling speed (31-35.5% rpm) for 30 seconds to stabilize.

CAUTION

If tailpipe temperature exceeds 710°C, a hot start has occurred; immediately return throttle to the closed position. Cause of hot start should be ascertained and the possible resultant damage surveyed before a restart is attempted.

Note

Maximum tailpipe idling temperature is 565°C. However, during engine acceleration from stop to idle, the temperature may exceed this maximum somewhat but will return to normal as engine operation stabilizes.

h. Cycle engine start master switch to facilitate air-start procedure in event of flame-out while airborne.

i. Push in inverter circuit breaker.

WARNING

If the throttle is closed inadvertently during engine operation, there will be an immediate flame-out which will be impossible to catch, regardless of how quickly the throttle is reopened. DO NOT TRY TO REGAIN IGNITION BY REOPENING THE THROTTLE, since raw fuel will be sprayed out of the tailpipe and will create a dangerous condition. A standard start or relight procedure will be necessary to restart the engine.

STARTING PROCEDURE—J48-P-8 ENGINE.

a. Check that throttle is at CLOSED.

b. Battery switch at BAT. & GEN.

c. Set fuel master switch to ON—after a few seconds, check that fuel boost pressure warning light goes out, indicating that sufficient fuel pressure (6 psi minimum) has been built up.

Note

This is not a positive check of the boost pump. A malfunctioning pump may build up 6 psi fuel pressure with a closed fuel system and still not provide sufficient pressure under flow conditions.

d. Set engine start master switch to ON.

e. Set engine fuel system selector switch to RESET PRIMARY and hold for two seconds, then set to START & PRIMARY.

f. Set cranking switch to START and release.

g. As soon as the engine commences to crank, move the throttle outboard to START.

h. Hold the throttle in the START detent until the tachometer indicates 5-6% rpm.

i. At 5-6% rpm, move the throttle to the IDLE detent, being careful not to advance past IDLE.

j. Allow the engine to accelerate and idle for 30 seconds to stabilize (33-35% rpm).

Note

If trouble is encountered while attempting to start on START & PRIMARY, set fuel system selector switch to EMER. and attempt an engine start as outlined.

WARNING

If engine was started with engine fuel system selector switch set at EMER., make sure engine performance is checked with this switch set at START & PRIMARY.

CAUTION

If tailpipe temperature exceeds 780°C, a hot start has occured: IMMEDIATELY close throttle. Cause of hot start should be ascertained and the possible resultant damage should be surveyed before restart is attempted.

Note

Maximum tailpipe idling temperature is 565°C. However, during engine acceleration from start to idling rpm, the temperature may exceed this maximum somewhat but will return to normal as engine operation stabilizes.

k. Cycle engine start master switch to facilitate air-start procedure in event of flame-out while airborne.

l. Push in inverter circuit breaker.

WARNING

If the throttle is closed inadvertently during engine operation, there will be an immediate flame-out which will be impossible to catch, regardless of how quickly the throttle is re-opened. DO NOT TRY TO REGAIN IGNITION BY REOPENING THE THROTTLE, since raw fuel will be sprayed out of the tailpipe and will create a dangerous condition. A standard start or relight procedure will be necessary to restart the engine.

FALSE STARTS.

If light-up does not occur within 15 seconds after advancing throttle to IDLE, or if engine fails to accelerate to idle rpm within approximately one minute after light-up, proceed as follows:

a. Set throttle to CLOSED.

b. Set engine start master switch to OFF.

c. When engine stops rotating, set fuel master switch to OFF and investigate.

CAUTION

Never set fuel master switch to OFF until engine stops rotating.

CLEARING ENGINE.

a. Cycle engine start master switch to ON, and fuel master switch to ON.

b. Set cranking switch to START and release.

c. Operate engine at 8-10% rpm for 30 seconds to clear engine.

d. Set engine start master switch to OFF to discontinue clearing procedure.

CAUTION

Do not set cranking switch to START until engine has stopped rotating.

HOT STARTS.

A hot start is a condition occurring during a starting attempt wherein the tailpipe temperature soars above normal and flaming or torching occurs at the tailpipe exit. It may be caused by a malfunctioning of the fuel metering system or by a faulty starting procedure, but it is the direct result of one condition—an excess of fuel in the combustion chamber. Use only the approved starting procedure for starting the engine. When a hot start occurs, observe the following procedure:

a. Set throttle to CLOSED.

b. When engine has stopped rotating, set fuel master switch to OFF.

c. Clear engine, using procedure given in Clearing Engine paragraph.

d. After investigation, if it has been determined that the hot start was caused by other than a system malfunction, attempt a restart.

Note

For procedure to be followed in case of fire, see Fire paragraph, Section III.

RUNAWAY STARTS.

A runaway start is a rare occurence, but can be encountered if a fuel control malfunctions, or if a section of the throttle linkage is disconnected. In such a case, the engine must be stopped by cutting off the fuel supply from the tanks to the engine. This is done by setting the fuel master switch to OFF (battery switch must be set at BAT. or BAT. & GEN.).

ENGINE GROUND OPERATION.

WARM-UP.

a. Allow engine to run at idling speed (33-35%) for approximately 30 seconds, as a normal stabilizing period.

b. With throttle set at IDLE, check the following instruments:

Tachometer—33-35%.

Tailpipe temperature indicator—565°C maximum.

Oil pressure—10 psi minimum, 15-25 psi normal.

Note

Prolonged idling at low rpm may result in excessive tailpipe temperature. If this occurs, operate at 55-65% for a short time. This rpm gives the coolest ground operating temperature. At periodic engine check time (after overhaul, etc), or if engine malfunctioning is suspected, an acceleration check may be made as follows: open throttle abruptly and during acceleration to 100% note tailpipe temperature—it should not exceed 800°C (acceleration limit). This maximum temperature is not to be maintained for more than five seconds. Time to accelerate from idle to 100% rpm should be 9-14 seconds.

c. Set engine fuel system selector switch to START & EMER. (J48-P-6A) or EMER. (J48-P-8). Amber emergency fuel system warning light will glow.

d. Plane captain conducts fuel integrity check.

e. Check operation of engine fuel boost pump by depressing boost pump cut-off switch for five seconds. If low fuel boost pressure light glows, the pump is not operating. Shut down and investigate.

f. Upon completion of fuel integrity check, set fuel system selector switch momentarily to RESET PRIMARY and then to PRIMARY (J48-P-6A) or START & PRIMARY (J48-P-8). Amber light will go out.

g. Advance throttle to 80% rpm and check cabin pressurization:

Set air conditioning system on-off switch to ON.

Close canopy.

Check cabin altimeter—decreasing.

Make sniff test for fuel leaks.

Plane captain makes high pressure fuel integrity check.

PRE-TAKE-OFF CHECK.

a. Hold airplane with brakes and advance throttle to full open. Check tailpipe temperature, 800°C or less, on acceleration. 100% rpm should be indicated.

b. Retard throttle to 55-60% rpm. Set engine fuel system selector switch to START & EMER. (J48-P-6A) or EMER. (J48-P-8).

c. Advance throttle smoothly to full open (100% rpm normal, 101.5% rpm maximum).

Note

With the fuel system selector switch at START & EMER. (J48-P-6A) or EMER. (J48-P-8), engine rpm may not reach 100% on cold days.

d. Retard throttle to 55-60% rpm. Set fuel system selector switch to RESET PRIMARY momentarily, then to PRIMARY (J48-P-6A) or START & PRIMARY (J48-P-8). Amber light will go out.

e. Ready for take-off:

Advance throttle to 100% rpm.

Check tailpipe temperature—755°C (J48-P-6A); 780°C (J48-P-8).

Check for 100% rpm.

Note

When operating in an ambient air temperature of −10°C (15°F) or lower at sea level, the tailpipe temperature indication may exceed the tailpipe temperature limitation during take-off at 100% rpm. This higher temperature reading is caused by a tendency of the engine during low ambient temperature conditions to run at a somewhat higher tailpipe temperature and also by indicator inaccuracy due to exposure to the cold. If, during a take-off at 100% rpm on a cold day, the tailpipe temperature appears to exceed this limit (after allowing for possible indicator error), engine rpm should be reduced so that the tailpipe temperature falls within the limitation. The resultant thrust of this reduced rpm will still be greater than or equal to that obtainable under standard day, 15°C (59°F), conditions when at 100% rpm.

CAUTION

Do not hold 100% rpm more than one minute during ground operation.

GROUND TESTS.

HYDRAULIC SYSTEM.

a. Check that hydraulic system pressure gage indicates 1400-1500 psi.

b. Cycle seat, adjust as desired and check for locked (adjust rudder pedals to conform to seat position).

c. Spread and lock wings (check indicators).

d. Check flaps and slats operation—slats fully extended when flaps are lowered.

e. Check flaperon operation.

f. Check speed brakes operation.

FLAPERETTE OPERATIONAL CHECK.[1]

With engine rpm at 60%, select flaperette EMER. ON & TEST position on flaperette system switch. After 10 seconds, operate lateral control system rapidly with simultaneous speed brake extension. The flaperettes can then be seen to move momentarily ahead of the flaperons. Return switch to AUTO position after check.

FLAPERETTE OPERATIONAL CHECK.[2]

With engine running, set flaperette system switch to EMER. ON & TEST position. Check that flaperettes deflect up in response to control stick movement. Return switch to AUTO position after check.

LONGITUDINAL CONTROL SYSTEM CHECK— INTERIM MODIFICATION.

a. Raise flaps.

b. Set longitudinal control system selector switch to HYDRAULIC.

c. Check that stabilizer hydraulic power emergency shut-off switch is set at the guarded position.

Note

Whenever the stabilizer hydraulic power emergency shut-off switch is set to its emergency position, the power operated stabilizer control system will be inoperative until the emergency shut-off valve is reset by the ground crew.

d. Trim stabilizer full airplane nose-down (1-3/4 degrees), using stick grip trim switch (neutral stick, zero stick force).

e. Trim stabilizer 0 degrees (neutral stick, zero stick force).

f. Move stick fore and aft full throw and check stabilizer motion in mirror and on the absolute stabilizer position indicator on instrument panel.

Note

The stabilizer travel limits will be 4-1/2 degrees airplane nose-down for full forward stick and 6-1/2 degrees for full back stick with 0 degrees stabilizer trim setting.

g. Retrim stabilizer to the airplane full nose-up trim position (6 degrees) (neutral stick, zero stick force).

h. Move stick fore and aft full throw and check stabilizer motion in mirror and on the absolute stabilizer position indicator on instrument panel.

Note

The stabilizer travel limits will be about 1-1/2 degrees airplane nose-up for full forward stick and 6-1/2 degrees airplane nose-up for full back stick with 6 degrees airplane nose-up trim.

i. Set longitudinal control system selector switch to MANUAL.

j. Retrim stabilizer to 3-1/2 degrees airplane nose-up (take-off setting).

WARNING

Retrimming must be done with the longitudinal control system set in manual.

k. Check operation of elevators for full throw.

l. Set longitudinal control system selector switch to HYDRAULIC.

m. Lower flaps and check for shift of longitudinal control from hydraulic powered to manual control.

n. Set longitudinal control system selector switch to MANUAL for take-off.

o. Recheck stabilizer trim setting of 3-1/2 degrees airplane nose-up for take-off.

LONGITUDINAL CONTROL SYSTEM CHECK— FINAL PRODUCTION MODIFICATION.

a. Raise flaps.

b. Set longitudinal control system selector switch to HYDRAULIC.

c. Check that emergency manual control knob is in up (normal) position.

d. Check that stabilizer trim selection switch is at NORM.

e. Trim stabilizer full airplane nose-down (1-3/4

[1] F9F-6 airplanes ser No. 127359 through 128279 and F9F-6P airplanes ser No. 127473 through 134452.

[2] F9F-6 airplanes ser No. 127327 through 127358, 128280 through 128294, 131029 and subsequent and F9F-6P airplanes ser No. 134453 and subsequent.

degrees), using stick grip trim switch (neutral stick, zero stick force).

f. Trim stabilizer 0 degrees using stick grip trim switch (neutral stick, zero stick force).

g. Move stick fore and aft full throw and check stabilizer motion in mirror and on the absolute stabilizer position indicator on instrument panel.

Note

The stabilizer travel limits will be 4-1/2 degrees airplane nose-down for full forward stick and 6-1/2 degrees airplane nose-up for full aft stick with 0 degrees stabilizer trim setting.

h. Retrim stabilizer to the airplane full nose-up trim position (2-1/2 degrees) (neutral stick, zero stick force).

i. Set stabilizer electrical trim selection switch to EMERG. and, using emergency trim switch, run the stabilizer to the airplane full nose-up trim position (6 degrees) (neutral stick, zero stick force).

j. Move stick fore and aft full throw and check stabilizer motion in mirror and on the absolute stabilizer position indicator on instrument panel.

Note

The stabilizer travel limits will be about 1-1/2 degrees airplane nose-up for full forward stick and 6-1/2 degrees airplane nose-up for full aft stick with 6 degrees airplane nose-up trim.

k. Holding neutral stick, lower flaps and check for shift of longitudinal control to manual.

l. Reset stabilizer electrical trim selection switch to NORM., retaining the 6 degrees airplane nose-up trim setting.

m. Raise the flaps and note that the stabilizer automatically returns to 2-1/2 degrees airplane nose-up.

n. Set longitudinal control system selector switch to MANUAL.

o. Check operation of elevators for full throw.

p. Lower flaps.

q. Retrim stabilizer for take-off (3-1/2 degrees airplane nose-up, or as desired).

ELECTRICAL SYSTEM.

GENERATOR CHECK.

With engine idling and external power disconnected, advance throttle and check voltmeter and generator warning light. Voltmeter should indicate 27.5-28.5 volts and generator warning light should remain out.

LIGHTS.

Check the operation of the following lights:

a. All press-to-test lights.

b. Fast erecting gyro horizon warning light.

Note

Warning light cover may be removed for daytime operation.

c. Fire warning lights. Check circuit continuity with test switch.

RADIO CHECK.

Check communication and navigational equipment.

FLIGHT INSTRUMENTS.

Set and align G-2 compass so that operation may be checked while taxiing. Uncage attitude gyro.

YAW DAMPER CHECK—F9F-6 AIRPLANES.

The Lear yaw damper system can best be checked while taxiing. Slow turns will give a rudder reaction opposing the turn.

AUTOMATIC PILOT—F9F-6P AIRPLANES.

Check auto pilot—OFF. See Section IV.

TAXIING.

Taxiing is facilitated by the excellent forward visibility. Approximately 50% rpm is required to start rolling, but a throttle setting slightly above idle is sufficient to maintain taxi speed. The nose wheel is swivelling and self-centering but is not steerable; use brakes to steer the airplane. In a crosswind, raising the downwind flaperon helps to reduce the weather cocking tendency. To stop, retard throttle and apply brakes evenly. For carrier operations, the brakes are adequate for spotting, but use of a nose wheel steering bar and/or main wheel catapult chock simplifies spotting operations. Keep the nose wheel on the deck, especially when using the nose wheel steering bar. Special care must be used when braking on a wet deck. Avoid using the brakes while crossing the barrier. When taxiing crosswind or on a carrier, keep the wings spread whenever possible.

Restrict taxi time to an absolute minimum, as fuel consumption during ground operation is exceedingly high. Fuel flow while taxiing is approximately 20-25 pounds per minute. Each minute of taxiing reduces range by five miles.

BEFORE TAKE-OFF.

PREFLIGHT AIRCRAFT CHECK.

a. Controls unlocked and surfaces move freely.

b. Attitude and directional gyros set as desired. Check operation.

c. Longitudinal control system switch set to MANUAL.

d. Stabilizer electric power switch set to NORM.

e. Speed brakes UP.

f. Yaw damper ON (F9F-6 Airplanes).

g. Auto pilot OFF (F9F-6P Airplanes).

h. Check "G" suit system. Depress button on "G" suit control valve.

i. Set cabin air conditioning system on-off switch to ON (TEMP) and regulate system for desired temperature by means of increase-decrease switch.

Figure 2-4. Take-off Check List

Note

On humid days and for all catapult take-offs, set the air conditioning system to OFF (RAM AIR) and regulate the system for the desired amount of ram air by means of the increase-decrease switch. Climb to 5000 feet before turning off the ram air. This will prevent fog formation in the cabin.

j. Set cabin air vent controls as desired.

Note

The following items (steps k. through v.) are those of the take-off check list.

k. Check flying tail.

l. Stabilizer (trim airplane)—nose up.

3.5°—Flying tail airplanes.

5.5°—Spring tab elevator airplanes.

WARNING

On flying tail airplanes, set stabilizer trim with the longitudinal control system selector switch at MANUAL. When operating with this switch set at HYDRAULIC, the absolute stabilizer position indicator denotes trim position only when the control stick is in neutral and no fore or aft stick force is being applied.

m. Wing and rudder trim—neutral.

n. Hydraulic pressure control lever—ON.

o. Wings locked.

p. Flaps down.

q. Fuel.

r. Fuel system selector switch.

PRIMARY	J48-P-6A engine
START & PRIMARY	J48-P-8 engine

s. Sliding nose—locked.

t. Oxygen—ON.

u. Shoulder harness (tight and locked).

v. Seat cycled and locked (positioned as desired).[1]

Note

For catapult take-offs, seat position should be such that pilot's head rests firmly against head-rest.

w. Taxi out into take-off position, heading airplane straight down runway with nose wheel straight.

TAKE-OFF.

FIELD TAKE-OFF.

Operational experience has shown that a most critical item of take-off performance is the ability to climb over an obstacle under conditions of high altitude and hot air temperature. Therefore, the distances given in figures A-1 and A-2 of Appendix I, this book, are based on the normal take-off technique described here, which results in minimum obstacle clearance distance for a flaps-down take-off, BUT NOT A MINIMUM GROUND RUN.

The normal lift-off airspeed results in greater margin above stalls, better handling qualities, and better performance.

a. Take-off check list completed (flaps down).

b. Hold brakes and advance throttle.

c. With engine at take-off rpm, and within limit tailpipe temperature, release brakes and begin take-off run.

d. Maintain direction with flaperons and with as little brake as possible until rudder control becomes effective (60-70 knots IAS). A raised flaperon adds drag to assist braking on that side. Excessive use of brakes or flaperons will increase the take-off distance. Sharp application of brake at speeds faster than 50 knots may result in a blown tire.

[1] F9F-6 airplanes ser No. 126257 through 127442 and F9F-6P airplanes ser No. 127473 through 127480.

THRUST AND DRAG DURING TAKE-OFF

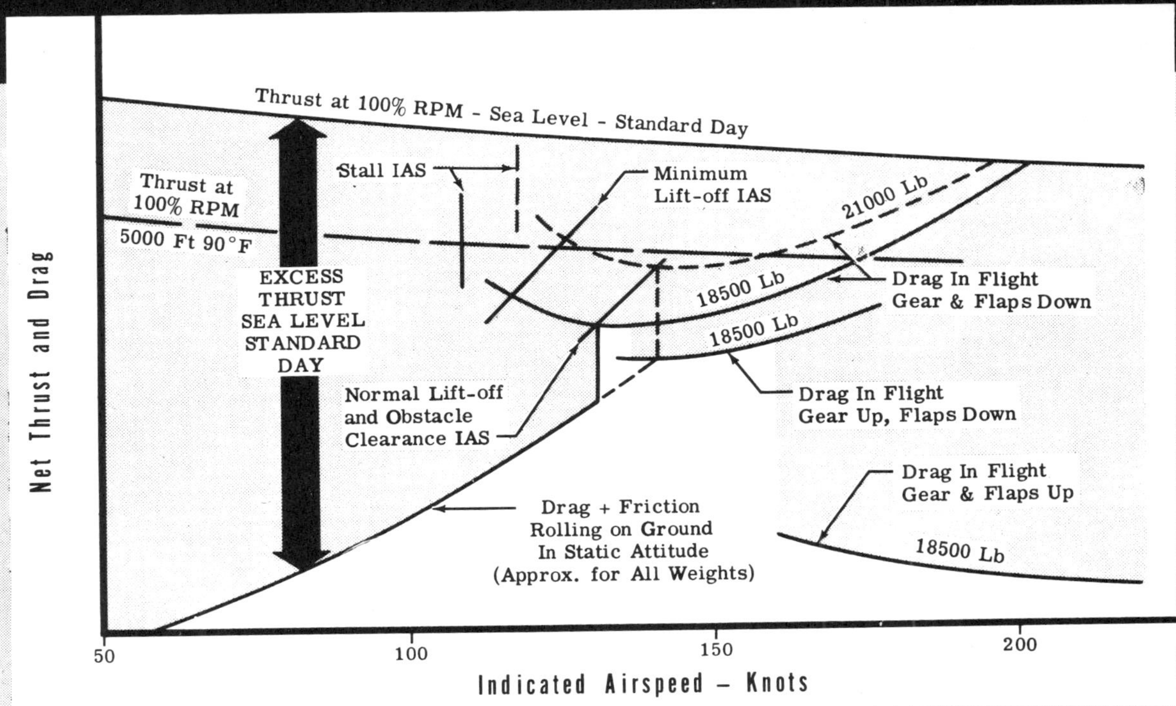

Note

1. $\frac{\text{Excess Thrust}}{\text{Weight}}$ = Forward Acceleration in g
2. $\frac{\text{Obstacle Ht x Wt}}{\text{Excess Thrust}}$ = Climb Distance over Obstacle in Feet
3. Note the advantage of keeping the nose down until normal lift-off IAS is reached and the double advantage of reducing take-off weight when thrust is low.

Figure 2-4A. Thrust and Drag During Take-off

e. At the recommended stabilizer trim setting (3-1/2 degrees airplane nose-up for flying tail airplanes; 5-1/2 degrees airplane nose-up for spring tab elevator airplanes), apply some push force on the stick to keep the nose wheel firmly on the ground. This reduces drag and weather-cocking tendency. At low fuel loadings (aft cg), greater push force will be required.

f. At the normal lift-off IAS given in Table II-I and in figures A-1 and A-2, raise the nose smoothly until the airplane lifts off.

g. Landing gear up when definitely airborne (retraction time about 10 seconds) and when remaining runway length is insufficient for landing in the event of an emergency.

h. For optimum performance to clear an obstacle, pull up smoothly so that airspeed does not increase. This will result in the optimum climb angle. The angle will be steepest when the landing gear is retracted, at light weights, and at low field altitudes and cold air temperatures.

DON'T LIFT NOSE WHEEL TOO SOON!

Note

At high field altitudes and hot air temperatures, the available thrust is low and the true speed (corresponding to the fixed IAS) to which the airplane must be accelerated is high. Thus, the time and distance required for take-off are increased. Be patient. Do not raise the nose until the normal lift-off IAS is reached.

i. After clearing obstacle, level off to accelerate to retraction speed. Raise flaps above 150 knots IAS for no stores configuration. With tanks or other heavy external stores, raise flaps at 160 knots IAS and 500 feet altitude. The acceleration to 160 knots IAS will be slow with stores installed. Note that minimum power-off glide IAS (flaps down) is 160 knots.

j. Close canopy.

k. For climbing speed schedule, see Climb Curves in Appendix I (380 knots IAS at sea level for no stores).

l. Whenever wing tanks are filled with more than 60 gallons (full fuel loading is 78 gallons per tank), set wing tank selector switches outboard to SELECT WING TANKS. If this is not done, decreased external pressure, due to increasing altitude, will cause the internal tank pressure to actuate a pressure relief valve and permit wing tank fuel to discharge through the ram air duct. Setting the wing tank selector switches to SELECT WING TANKS will prevent this loss of fuel. After approximately two minutes (engine fuel consumption will have decreased wing tank fuel loading by 18 gallons per tank), the switches may be returned to the center (off) position. Fuel will not be lost when 60 gallons or less remain in each wing tank.

m. When at a safe altitude (5000 feet above terrain recommended), trim to neutral stick, zero stick force, and shift longitudinal control system to hydraulic power.

WARNING

For familiarization flights, it is recommended that the shift be made at 15000 feet.

Note

There may be a small trim change when shifting from one system to the other. It is recommended that the shift be made at speeds of 250 to 350 knots IAS.

n. For take-off emergencies, see Section III.

Note

If during take-off, thrust and rpm should drop slowly indicating impending fuel system failure, immediately set fuel system selector switch to START & EMER. (J48-P-6A) or EMER. (J48-P-8).

TABLE II-I.

LIFT-OFF AIRSPEEDS

Airplane Weight (pounds)	*Normal Lift-off IAS (knots)*	*Minimum Lift-off IAS (knots)*	*Obstacle Clearance IAS (knots)*
22000	141	127	141
20000	133	120	133
18000	125	112	125
16000	117	104	117

MINIMUM GROUND RUN.

For minimum ground run, follow the procedure given in Field Take-off paragraph, except raise the nose smoothly at the minimum lift-off IAS (Table II-I). A relatively high angle of attack will be required. Immediately after lift-off, level out, retract landing gear, accelerate to the obstacle clearance IAS (Table II-I) and then pull up smoothly to keep airspeed constant until obstacles are cleared.

The ground run will be approximately 75% of the ground run distances shown in figures A-1 and A-2 of Appendix I, this book, and the total obstacle clearance distance will be approximately 120 to 170% of the total distance shown in these figures (the larger value applying to conditions of high altitude and hot temperature). This is because the airplane's acceleration in the air is slower than its acceleration on the ground, because of the increased drag accompanying the increased angle of attack required to maintain flight.

CATAPULT TAKE-OFF.

Note

For recommended catapult end speeds, see figure 6-6. For catapult pressures and required wind over the deck, see applicable Catapult Bulletins.

a. Take-off check list completed.

b. Tail skid control switch in guarded position to assure that tail skid will retract when throttle is advanced, to avoid possible contact with catapult shuttle.

c. As soon as airplane has been tensioned on catapult, release brakes and keep feet clear of brake pedals.

d. Lift catapult grip and advance throttle to OPEN. Grasp throttle and catapult grip together firmly.

e. Assume a comfortable position against back of seat and headrest.

f. When engine reaches take-off rpm, give prescribed catapulting signal.

g. Hold control stick firmly, but not rigidly, in approximately the neutral elevator position.

h. At end of catapult power stroke, the pilot will be able to break away from the headrest and fly the airplane off the bow in approximately the attitude it has achieved at the end of the power stroke. (See Catapult Take-off Characteristics paragraph in Section VI.)

i. After catapulting, turn away from carrier path.

CAUTION

Under conditions of marginal thrust, the turn must be very shallow to avoid losing airspeed. Do not increase altitude in the turn.

j. Raise landing gear.

k. Close canopy.

l. Flaps up above 150 knots IAS for no stores configuration. With tanks or other external stores, retract flaps at 160 IAS and 500 feet altitude. Best acceleration is obtained by not increasing altitude. Acceleration to 160 knots IAS will be slow with stores installed.

m. Accelerate to best climbing speed (390 knots approximately). See Climb Curves in Appendix I.

n. When at a safe altitude (5000 feet recommended), trim to neutral stick, zero stick force and set longitudinal control system selector switch to HYDRAULIC.

Note

There may be a small trim change when shifting from one system to the other. It is recommended that the shift be made at speeds of 250 to 350 knots IAS. For familiarization flights, it is recommended that the shift be made within this speed range and at 15000 feet.

CLIMB.

Climb at 100% rpm or the rpm for tailpipe temperature limits. (See Section V, figure 5-2 for engine operating limits.) The recommended climb speed at sea level for the airplane with no external stores is 390 knots IAS. Other climb data are given in Appendix I. After reaching a safe altitude (5000 feet recommended) set longitudinal control system selector switch to HYDRAULIC.

FLIGHT CHARACTERISTICS.

See Section VI.

SYSTEMS OPERATION.

See Section VII.

DESCENT.

Utilize speed brakes to decelerate. Use of idle rpm (throttle at IDLE setting) during descent will conserve fuel.

Note

Due to engine compressor blade resonant vibration peak, it is recommended that the rpm range of 68-76% be used as little as possible.

Two standardized descent schedules are given in Appendix I in figures A-12 and A-13. Descents with the fuel low level warning light on must be made in accordance with the procedure given in Appendix I, Description of Charts and Tables paragraph, step h.

PRE-TRAFFIC PATTERN CHECK LIST.

a. Armament master switch—OFF; gun switches—SAFE.

b. Hydraulic pressure control lever—HYDRAULIC PRESSURE-ON.

c. Wing tanks—empty.

Note

When airplane is decelerating, inertia forces will prevent fuel from being dumped from wing tanks.

d. Set wing tanks dump switch to NORMAL.

e. Check communication equipment.

f. Auto pilot—OFF (F9F-6P).

g. Yaw damper—ON (F9F-6).

h. Trim to neutral stick, zero stick force, and shift longitudinal control system to manual (set longitudinal control system selector switch to MANUAL) (5000 feet altitude recommended).

Note

There may be a small trim change when shifting from one system to the other. It is recommended that the shift be made at speeds of 250 to 350 knots IAS. For familiarization flights, it is recommended that the shift be made within this speed range and at 15000 feet.

i. Airplane trim—as required.

TRAFFIC PATTERN CHECK LIST.

a. Lower landing gear.

Note

Both the landing gear and the landing gear control lever must be checked down AND LOCKED.

b. Arresting hook—extend for carrier landing.

c. Flaps down.

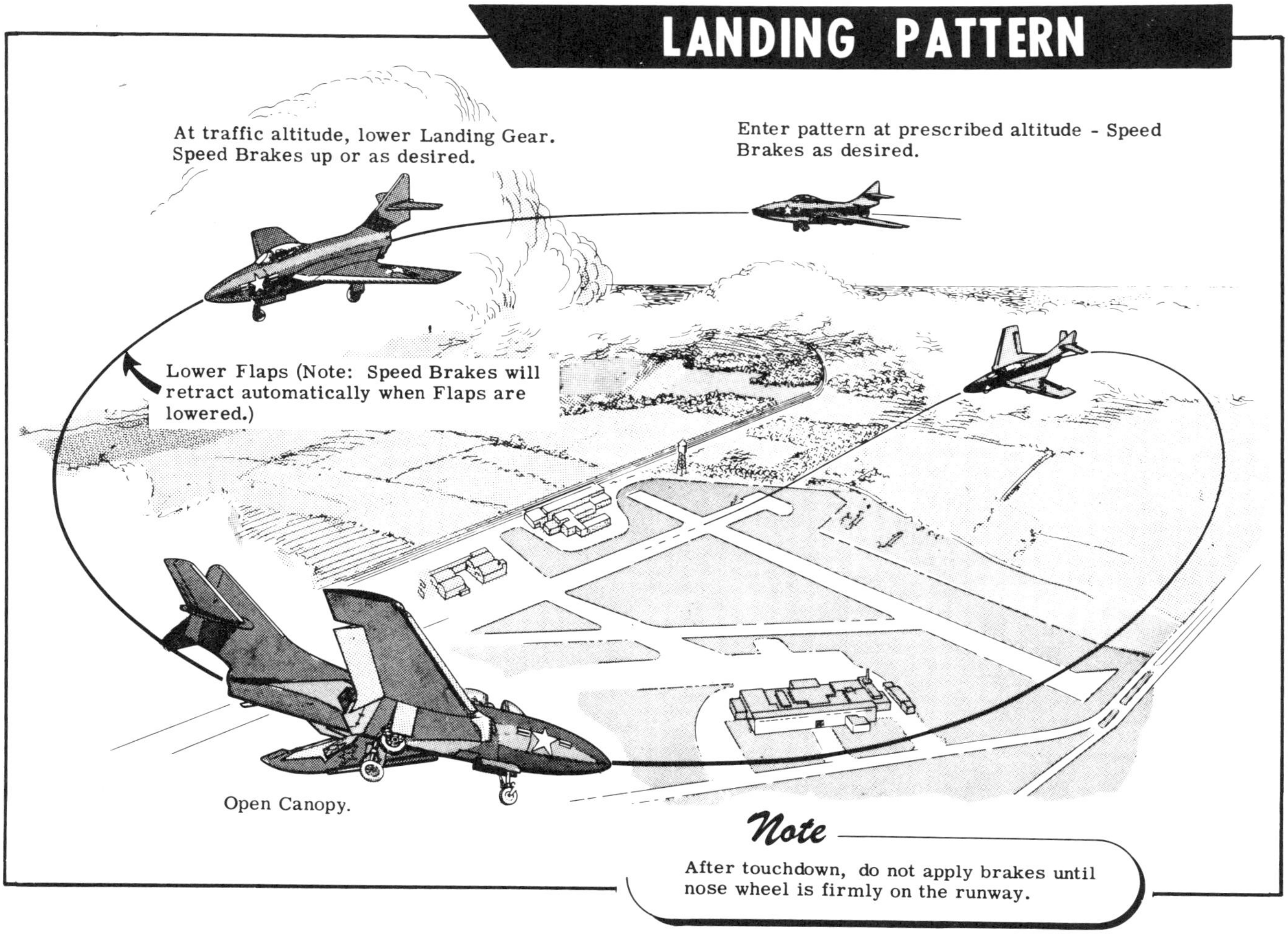

Figure 2-5. Landing Pattern

Note

If speed brakes have been used to decelerate, they will retract automatically before the flaps extend (if the speed brakes emergency landing override switch is in the normal position).

d. Depress brake pedals—check "feel".

e. Check fuel quantity. See Appendix I for Fouled Deck Endurance data.

f. Shoulder harness and seat belt—locked and tightened.

g. Canopy—open for field landings. For carrier landings, follow squadron practice.

FIELD LANDING.

See Section V and technical orders for maximum allowable landing weights. For normal field landings, make final approach at 130% of power-off stall speed and with wheels and flaps extended (see Table II-II). Avoid high rates of descent on final approach, as airplane's response to landing flare-out is slower than that of a straight wing airplane. For landing pattern diagram, see figure 2-5. For emergency procedure, see Section III.

CAUTION

Because of the slower accelerating characteristics of the jet engine, landing traffic pattern should be made to permit engine operation above 65% rpm.

Note

Limit engine operation between 68-76% rpm to as short a time as possible.

Touch-down should be made at about 110% of the power-off stall speed (see Table II-II).

For normal roll-out:

a. Keep flaps down.

b. Let nose wheel come down.

c. Use brakes. For extended tire life use only as much braking as necessary.

For minimum landing roll:

a. Get nose wheel on runway as soon as possible after normal touch-down.

b. Apply brakes gradually at first, then apply harder as ground speed decreases.

c. Set throttle to CLOSED.

TABLE II-II

NORMAL FIELD LANDING AIRSPEEDS

Airplane Weight (pounds)	*Normal Approach IAS (knots)*	*Normal Touch-down IAS (knots)*
18000	142	120
16000	134	113
14000	125	105

CAUTION

On F9F-6P airplanes, generator cooling is marginal at indicated airspeeds below 185 knots at altitudes less than 5000 feet. Avoid unnecessary operation under these conditions.

CROSS-WIND LANDING.

Cross-wind approach and touch-down techniques used for straight wing airplanes are applicable. A successful technique during roll-out is to get the nose wheel on the runway immediately after landing and, as soon as it is determined that a go-around is not required, raise the flaps. This reduces the weather cocking tendency and both brakes may then be used simultaneously to stop the airplane. Raising the down-wind flaperon is also a useful supplement to rudder control for keeping the airplane straight during roll-out or taxiing.

CARRIER LANDING.

Pre-traffic pattern and traffic pattern check list completed (see figure 2-7). For carrier landings with no external stores, the recommended approach speed is 120 knots ± 2 knots. (For required wind over deck, see figure 6-7.)

Engine rpm required in the approach will vary with gross weight from about 85% rpm at 15500 pounds to about 80% rpm at 13500 pounds. It is usually necessary to add about 2% rpm in the steepest part of the approach turn to avoid decelerating and then to reduce the power approximately the same amount on rolling out in the groove to avoid accelerating. The approach in the groove will probably be of approximately 5-10 seconds duration. Therefore, major corrections of airspeed and altitude must be made before getting in the groove. The airplane has a somewhat higher angle of attack in the carrier approach than does a straight wing airplane. For this reason, the induced drag is considerably higher and slight reductions in thrust result in rather rapid deceleration. Also for the same reason, comparatively large increases in thrust must be made to increase the speed of the airplane, particularly when correcting a low or slow landing approach. Do not take off too much power when in the groove. After the cut, establish a rate of

CARRIER LANDING PATTERN

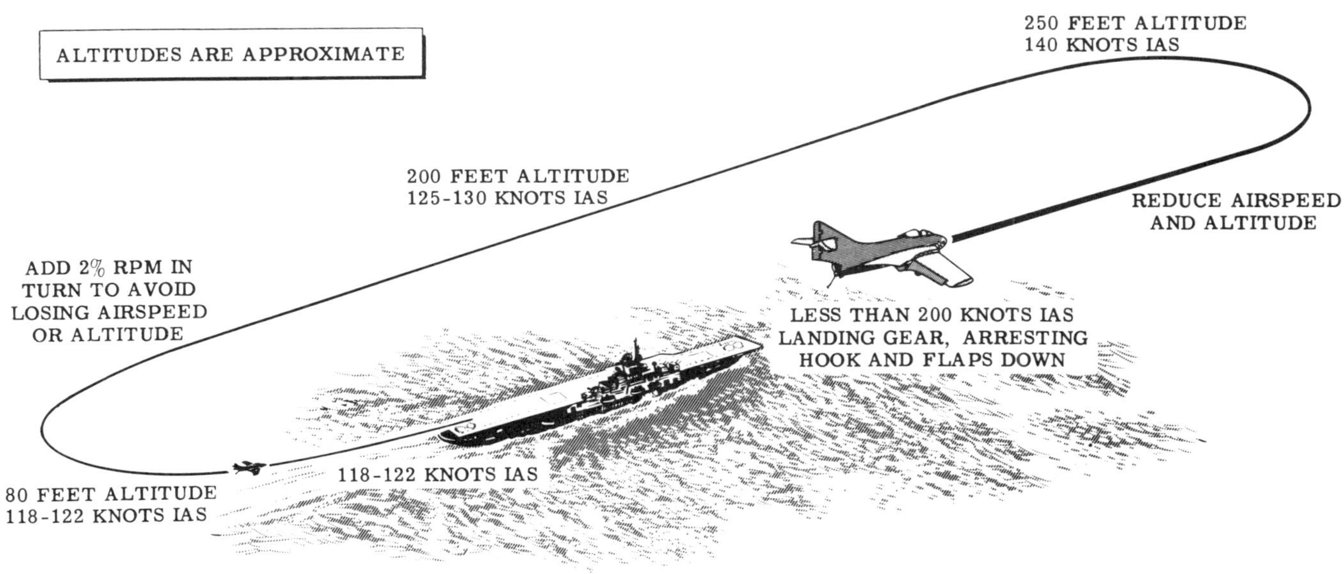

Figure 2-6. Carrier Landing Pattern

Figure 2-7. Landing Check List

descent and maintain the resulting attitude; do not attempt to flare. Land straight ahead; do not attempt to line up with the deck center line. It may be necessary to rotate the airplane slightly forward immediately before touch-down so that when the main wheels touch, the nose wheel and tail skid are equidistant from the deck. After touch-down, keep the nose wheel on the deck.

WAVE-OFF.

A swept wing airplane is a little slower to accelerate on a wave-off than a straight wing airplane, because the drag, in the landing configuration, is greater for the swept wing. Therefore, be careful to take a wave-off soon enough. Also, a jet engine responds slower than a reciprocating engine. Maintaining at least 65% rpm while making field approach will assist in making a rapid wave-off. (See Section VII for engine accelerating times.)

a. Throttle—full forward.

b. Landing gear up when touching the runway is no longer a possibility.

c. Flaps up—above 150 knots IAS.

Note

Approximately 220 pounds of fuel will be required to go around again in a tight pattern.

AFTER LANDING.

a. At end of taxi run, raise flaps.

b. If tailpipe temperature is high from hard taxiing or stationary ground running, operate at 55-65% for one minute.

CAUTION

After wings are folded, it is recommended that wing fold jury struts be installed before the engine is shut down.

AFTER ARRESTING.

a. As airplane rolls back, advance throttle to 100% rpm (approximately 8 seconds required for engine to reach 100% rpm from idle).

b. Apply brakes to prevent airplane from rolling forward.

c. Retract flaps.

d. When "hook clear" signal is received, raise hook and release brakes.

e. Coordinate throttle and taxi speed so that brakes are not required while crossing the barrier (see Taxiing paragraph).

f. Fold wings if desired. If taxiing onto the deck-edge elevator, delay folding until after having stopped, aligned fore and aft, on the elevator. This will preclude the possibility of the airplane rolling over.

STOPPING ENGINE.

a. Retard throttle to CLOSED from a stabilized rpm.

Note

Check that engine runs down freely. Engine driven accessories will slow down the engine. Any undue friction should be noticeable.

b. After engine stops rotating, set fuel master switch to OFF.

c. After 10 seconds, set engine start master and battery switches to OFF.

BEFORE LEAVING AIRPLANE.

a. Set all switches to OFF.

b. Install surface controls lock (bridle) and make up seat belt over front of stick.

c. Close canopy, using handle on left side of fuselage, after leaving the cabin.

d. Be sure airplane is properly chocked.

NAVY

SECTION III

TABLE OF CONTENTS

SECTION III EMERGENCY PROCEDURES

ENGINE FAILURE.

DURING TAKE-OFF.

a. If power fails before leaving ground, retard throttle to CLOSED and use brakes.

b. If power fails after becoming airborne, but while there is still runway left, retard throttle to CLOSED, land, and use brakes. For minimum stopping distance, fly the airplane onto the ground, force the nose wheel on the ground with full forward stick, and begin using brakes. Apply the brakes gently at first, increasing pedal pressure as they take effect. Be careful in applying brakes to avoid blowing a tire, because braking effectiveness with blown tires is poor. For this same reason, the emergency air brake should not be used until speed is below at least 50 knots, since the air brake will blow both main tires if applied at too high a speed.

c. If power fails after becoming airborne, and there is no runway left, retard throttle to CLOSED, move landing gear control lever to UP, move wing flap lever to DOWN, and land straight ahead.

d. If the failure occurs above several hundred feet, push over, hold 160 knots, and follow procedure in step c., above. If time permits, an abbreviated emergency engine relight procedure may be tried as follows:

Fuel system selector switch to START & EMER. (J48-P-6 engine) or EMER. (J48-P-8 engine).

Throttle to IDLE.

Pull emergency igniter system control handle.

Open throttle as engine rpm increases through idle. (See Airstart paragraph, this section, for complete procedure.)

PARTIAL ENGINE FAILURE DURING TAKE-OFF.

a. If a partial loss of engine thrust occurs before leaving the ground, proceed as in step a. of Engine Failure During Take-Off paragraph.

b. If partial loss of engine thrust occurs after becoming airborne but while there is still runway left, proceed as in step b. of Engine Failure During Take-off paragraph.

c. If a partial loss of engine thrust occurs after becoming airborne and there is no runway left, IMMEDIATELY set fuel system selector switch to EMER. If switching to emergency fuel control is delayed until rpm has fallen off to less than 80%, it would be advisable, time permitting, to first retard throttle to IDLE, then to switch to the emergency fuel control and advance throttle smoothly to obtain thrust needed for flight.

ENGINE FAILURE IN FLIGHT.

Note

The information and procedures presented in this section are based on actual flight tests.

GENERAL.

Although sudden failure of the engine in flight may be

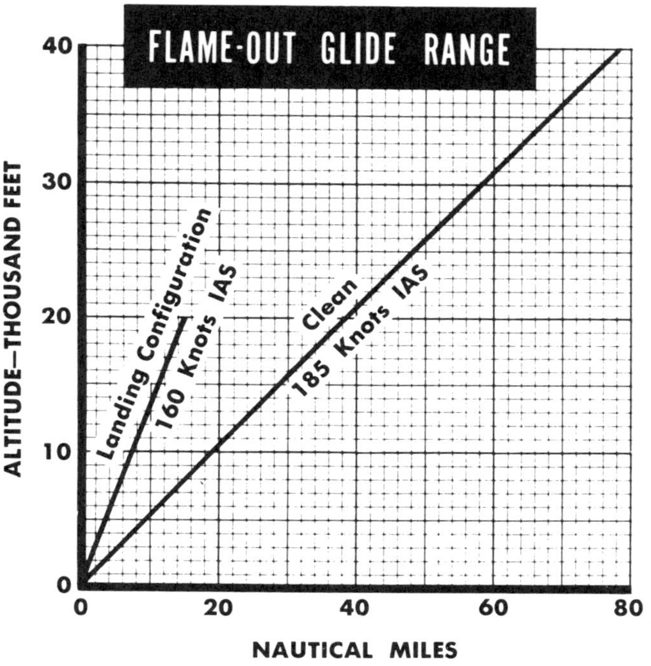

Figure 3-1. Flame-out Glide Range Curves

an alarming surprise to the pilot, the situation probably will not be critical immediately. There will be no change in the controllability of the airplane. If the engine has flamed out because of trouble in the primary fuel control, the chances of obtaining a relight in the emergency fuel control are excellent. If the engine failure is more serious and it cannot be restarted, a controlled let-down and satisfactory dead engine landing or ditching can probably be made without unusual difficulty. Immediately after an engine failure, the following conditions will be apparent to the pilot:

a. Loss of thrust, rapid dropping of engine rpm and tailpipe temperature.

b. Failure of the cabin pressurization system. The noise level in the cabin will become quieter and cabin altitude will increase to the airplane altitude.

c. Failure of the electrical circuits on the main bus of the electrical system as engine windmill rpm drops below generator cut-in speed. (Provided battery switch is in BAT. & GEN. position.)

A procedure of steps to be taken after an engine failure is included in this section; however, the following general information should be kept in mind:

a. For maximum gliding range, the airplane should be as clean as possible.

b. For maximum range in the clean configuration, maintain 185 knots IAS.

An indicated airspeed 10 knots higher or lower than 185 knots reduces gliding distance only two per cent, so strict adherence to this airspeed is not necessary.

The best approach speed in the landing configuration is 160 knots IAS. The use of any approach speed less than this will very likely result in a hard landing due to insufficient airspeed for flaring out.

The gliding distance in the clean and landing configurations is shown in figure 3-1. The rate of descent at various altitudes is shown in figure 3-2.

c. The windmilling engine will provide 1500 pounds of hydraulic pressure and sufficient capacity to operate the flying tail, flaperons, speed brakes, landing gear, and landing flaps. Hydraulic components operate faster on airplanes equipped with the flying tail than for those equipped with spring tab elevators because a larger capacity engine driven hydraulic pump is installed on flying tail airplanes. Times required to extend the landing gear and wing flaps for both configurations are shown in Table III-I.

d. Lateral control is satisfactory in an engine-off descent and landing for all airplane configurations.

On airplanes equipped with the flying tail (large capacity hydraulic pump), the windmilling engine will maintain adequate hydraulic pressure for normal flaperon control until the airspeed drops below 70 knots after touchdown.

On airplanes equipped with spring tab elevators, the windmilling engine will maintain adequate hydraulic pressure for normal flaperon control down to 140 knots IAS. Automatic switchover to flaperette control will occur instantly when hydraulic pressure drops below that necessary for flaperon control.

Note

For complete information on the emergency flaperette control system, see Emergency Flaperette Operation paragraph in this section.

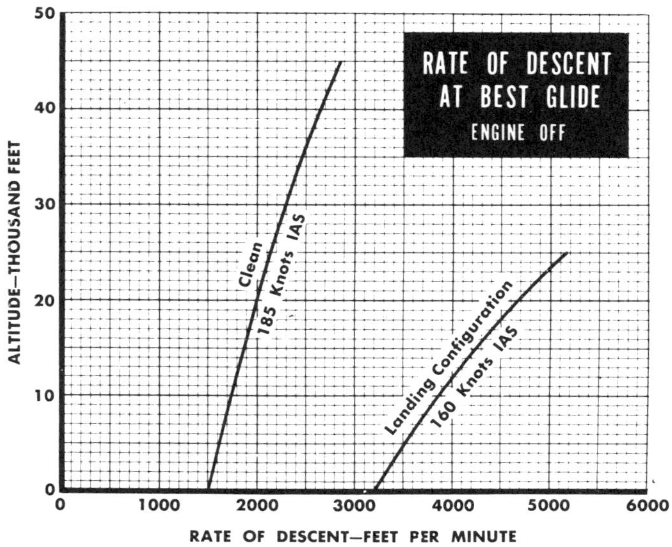

Figure 3-2. Rate of Descent at Best Glide Curves

TABLE III-I

TIMES REQUIRED FOR LANDING GEAR AND WING FLAP ACTUATION AIRPLANES EQUIPPED WITH FLYING TAILS OR SPRING TAB ELEVATORS

	Airplanes Equipped with Flying Tail	*Airplanes Equipped with Spring Tab Elevators*
To Extend Landing Gear 185 knots 10000—12000 ft		
Auxiliary Hydraulic Pump OFF	10—12 sec	35—45 sec
Auxiliary Hydraulic Pump ON	—	12 sec
To Extend Wing Flaps 160 knots 8500 ft		
Auxiliary Hydraulic Pump OFF	30 sec	50 sec
Auxiliary Hydraulic Pump ON	—	20 sec

Note

At higher altitudes, landing gear and wing flaps extend at faster rates because of increased engine windmill speed at same IAS.

On flying tail airplanes, landing gear and wing flap actuation times are increased if the flying tail is being used, therefore, the flying tail should be turned off when making an engine-off descent.

The landing gear may be extended by using the emergency air bottle. Extension time is five to six seconds. However, if landing gear is extended by use of the air bottle, it CANNOT be retracted. See Emergency Landing Gear Operation paragraph, this section.

e. With the engine windmilling at less than generator cut-in speed, all electrical circuits except those connected to the essential bus are automatically disconnected when the battery switch is in the BAT. & GEN. position and the landing gear control handle is UP. (For circuits connected to the essential bus, see figure 1-8.) If it is desired to operate a circuit on the main bus, such as the radio, auxiliary hydraulic pump, or exterior lights, the battery switch may be placed in the BAT. position. Placing the landing gear control handle in the DOWN position also energizes the main bus. With the battery switch in BAT., a perfect battery will last approximately 13 minutes. If the battery is conserved by pulling the circuit breakers of unnecessary systems, it will last approximately 25 minutes. If an airstart is to be attempted, it is best to use as little electrical power as possible to ensure that the battery power will be sufficient for the airstart. If an airstart is to be attempted, never set the battery switch to OFF as this cuts off the electrical fuel boost pump.

f. Canopy frosting may occur during engine-off descents. It will be the most severe when descending from high altitudes into an atmosphere of high humidity, The frosting can be kept to a minimum by opening the canopy slightly.

PROCEDURE.

The following procedure applies to the situation where the landing gear, wing flaps, and speed brakes are up when the engine fails and the airplane has at least several thousand feet of altitude over the terrain.

a. Retard throttle to CLOSED.

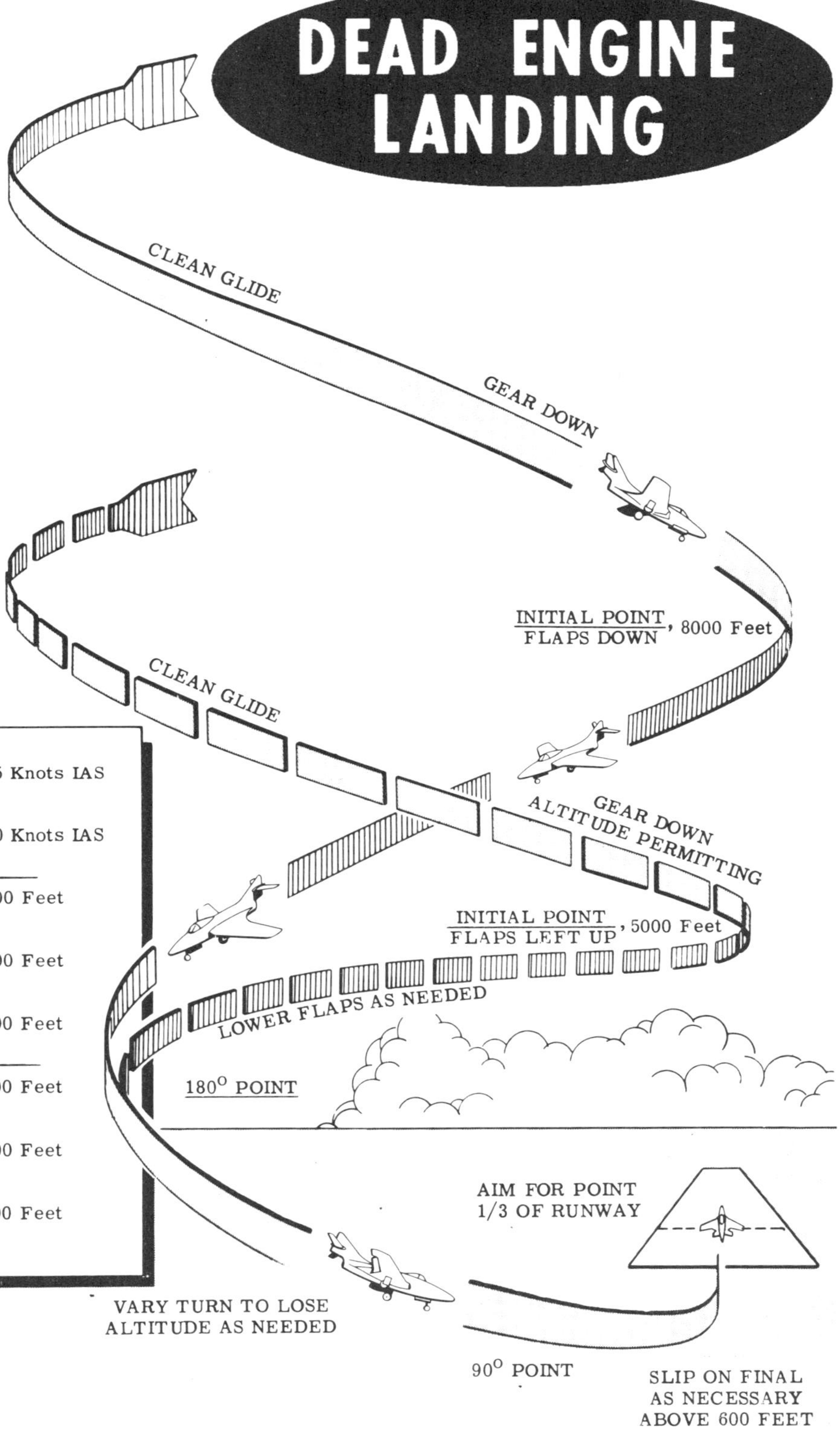

Best Glide Speed Clean Airplane	185 Knots IAS
Recommended Approach Speed Landing Condition	160 Knots IAS
Optimum Altitude at Initial Point Gear and Flaps Down	8000 Feet
Optimum Altitude 180° Point Gear and Flaps Down	4500 Feet
Optimum Altitude 90° Point Gear and Flaps Down	2500 Feet
Optimum Altitude at Initial Point Gear Down	6000 Feet
Minimum Altitude at Initial Point Gear Down	5000 Feet
Minimum Altitude 180° Point Gear Down	2500 Feet

Figure 3-3. Dead Engine Landing Pattern

b. Establish 185 knots IAS as soon as possible for maximum range glide. If speed is above 185 knots, pull up and climb until speed drops to 185 knots.

c. If at high altitude, check oxygen mask for tight fit after pressure surge following loss of cabin pressurization.

d. Dump wing tank fuel and jettison external stores, if carried.

Note

Under certain conditions, attempt a relight before dumping wing tank fuel in order not to jeopardize chances for a safe return if relight attempts are successful.

When airplane is decelerating inertia forces will prevent fuel from being dumped from wing tanks.

e. Check that battery switch is in BAT. & GEN. position to conserve battery. If power is desired for communications or for auxiliary hydraulic pump, set switch to BAT. and pull out circuit brakers to deenergize unnecessary equipment.

f. On airplanes equipped with the flying tail longitudinal control system, set longitudinal control system selector switch to MANUAL to reduce demand on hydraulic system.

g. Initiate airstart procedures when below 25000 feet, using normal relighting procedure. If normal relight procedures are unsuccessful, use emergency igniter relight procedure when below 18000 feet. If altitude is low and time is limited, use emergency igniter relight procedure first. (See Airstart paragraph, this section, for relight procedures.)

h. If all attempts to relight are unsuccessful and the situation occurs at night, during bad weather, or over mountainous terrain, eject from the airplane. (See Bail-Out paragraph, this section, for ejection procedure.)

i. If an airstart has not been achieved after repeated attempts, or by the time 10000 feet is reached, and the situation occurs during daylight hours under favorable conditions, all efforts should be turned toward making a successful landing.

j. If engine failure occurs below 10000 feet, primary emphasis should be on making a successful landing. Do not jeopardize chances of making a successful landing by overconcentration on relight attempts.

EMERGENCY LANDINGS.

FORCED LANDING—DEAD ENGINE.

PATTERN.

The recommended pattern for dead engine landings and simulated flame-out landings is shown in figure 3-3. The pattern consists of a 360 degree approach with two fixed check points. The initial point is defined as a position above and slightly to starboard of the intended landing point, with the airplane headed in the direction of the landing runway. The 180 degree point is the downwind position abeam of the intended landing point from which the turn onto base leg is made. Two possible patterns are presented because it is possible to make the approach with landing gear and flaps down all the way from the initial point, or to lower only the gear at the initial point and lower the flaps when needed later in the approach. If sufficient altitude is available, it is better to lower the flaps at, or before, the initial point, so that the flaps will extend fully and the rate of descent stabilize early in the approach. A constant rate of descent will make it easier for the pilot to judge his gliding angle and plan his approach. However, if there is insufficient altitude to make this type of approach, it is possible to make the approach with the flaps up and extend them when needed to hit the desired landing point. This type of approach is not unduly difficult, but more judgment must be exercised by the pilot during the approach. By following one of these patterns and using the pattern altitudes and airspeeds in figure 3-3, a pilot should be able to hit the desired landing point with relative ease. When using the recommended procedure for a dead engine landing, the pilot is required to exercise a minimum amount of judgment. This is beneficial in the event of a dead engine landing, because a pilot's judgment is likely to be impaired by the tenseness of the situation.

GENERAL PROCEDURE.

When a pilot is committed to a dead engine landing, either because of fuel exhaustion or failure to get an airstart after a flame-out, he should execute the following general procedure.

Establish the recommended clean gliding speed of 185 knots IAS and plan the let-down pattern so as to arrive over the initial point at the correct altitude for the type of approach to be made. The following steps should be started at sufficient altitude so as to be completed upon arriving over the initial point.

a. Retard throttle to CLOSED.

b. Set fuel master switch to OFF.

c. Set battery switch to BAT.

d. Move longitudinal control system selector switch to MANUAL.

e. Set auxiliary hydraulic pump to ON.

f. Move landing gear control lever to DOWN, if a safe landing is probable.

Note

If there is insufficient altitude to make the field with landing gear down, the landing gear should be left up and extended only when the field is definitely within reach.

The landing gear may be extended quickest by using the emergency landing gear air bottle. The extension time will be five to six seconds. Once extended by use of the air bottle, the landing gear CANNOT be retracted. See Emergency Landing Gear Operation paragraph, this section.

g. Set speed brake landing override switch to ON.

h. Lock and tighten shoulder harness. Remove personal connections (oxygen, "g" suit, radio leads).

i. Open canopy.

Extend flaps upon reaching the initial point, altitude permitting, and reduce speed to 160 knots IAS. Commence a left turn, banked 25 to 30 degrees. If the altitude over the initial point is higher than recommended, delay the turn slightly or decrease the bank so the airplane will arrive at the 180 degree position at the correct altitude. If the initial point is reached with less altitude than the 8000 feet recommended for the high 360 degree patterns, continue the approach with flaps up, and lower the flaps as needed to reach the 180 degree position at the recommended 4500 feet, or to intercept the flaps-down landing path. When a crosswind prevails, shallow or steepen the turn from the initial point to the 180 degree position if the crosswind is from the port or starboard, respectively, in order to have the correct distance at the 180 degree position. At the 180 degree position, check the altitude, angle of descent and position to determine if the same turn may be continued, or if a shallower or steeper bank should be used. At the 90 degree position, the airplane should be approximately 5000 feet down-wind of the landing point, and the altitude should be approximately 2500 feet with the flaps down. Plan the final turn in order to reach the straightaway with about 1000 feet of altitude. If the airplane is high, nose over and pick up additional speed to steepen the glide path. The speed brakes may be opened to further steepen the glide path.

Note

The longitudinal control system selector switch should be set at MANUAL and the speed brakes landing override switch must be set to its emergency position before lowering the speed brakes.

It is also possible to use a forward slip in order to lose some altitude, but this should not be carried any lower than about 600 feet, because of the high rate of descent it will cause, and the resultant increased time and altitude required to flare out for landing. Normally, commence the flare-out at about 200 feet above the runway. In the event of a steeper glide path, brought about by the use of speed brakes, slips, or nosing over, commence the flare-out as high as 400 feet above the runway. When beginning the flare-out at altitudes above 200 feet, exercise care not to flare out too abruptly, causing the airplane to run out of airspeed at too high an altitude. If, at any time in the approach up to the last 500 feet of altitude, it is seen that the airplane is undershooting, maintain airspeed at 160 knots and retract the flaps partially or fully until the final approach, or until there is enough altitude so that they may again be lowered. If the airspeed is maintained, no noticeable sinking will result from retraction of the flaps and the glide range will be extended considerably. If there is still insufficient altitude to make the field with gear down, retract the landing gear and effect a wheels-up landing. THE MOST IMPORTANT SINGLE THING TO REMEMBER ABOUT THE DEAD ENGINE LANDING APPROACH IS THAT UNDER NO CIRCUMSTANCES SHOULD THE AIRSPEED FALL BELOW 160 KNOTS. Because of the higher than normal altitudes used on the flame-out landing, there is a strong illusion that the landing will be overshot. This is likely to cause the pilot to lengthen out his pattern and undershoot. For this reason, a point one-third the way down the runway should be the intended landing point. Make a normal landing and roll-out. If the airplane is high and fast over the end of the runway and it appears that insufficient runway remains for a normal landing and roll-out, fly the airplane onto the ground fast, force the nose wheel on the ground with full forward stick and begin using brakes. Apply the brakes gently at first, increasing pressure as they take effect. Be careful in applying brakes to avoid blowing a tire, because braking effectiveness with blown tires is poor. For this same reason, the emergency air brake should not be used until speed is below 50 knots, since the air brake will blow both main tires if applied at too high a speed.

SIMULATED DEAD ENGINE FORCED LANDINGS.

By practicing simulated dead engine forced landings, a pilot will develop sufficient skill and confidence in the maneuver so that an actual power-off landing will present no undue hazard.

To simulate a flame-out, the following rpm, with speed brakes extended, are the equivalent of a dead engine:

ALTITUDE	RPM
30000—20000 feet	85%
20000—10000 feet	80%
Below 10000 feet, with landing gear and flaps down	65%

The longitudinal control system selector switch should be set at MANUAL when performing simulated dead engine forced landings.

Note

To keep speed brakes extended when flaps are down, the speed brakes landing override switch must be ON.

When practicing simulated dead engine landings, the landing patterns of figure 3-3 should be followed. The pilot may vary his altitudes from those recommended in figure 3-3 to determine the easiest flame-out approach for him.

In practicing simulated flame-out landings, the following precautions should be observed:

a. Normal field landing weight restrictions should be observed.

b. Before a pilot carries a simulated flame-out approach to a landing, he should make at least three approaches, taking a wave-off upon reaching 500 feet of altitude.

c. Until pilots become thoroughly skilled in the maneuver, simulated flame-out landings should be made touch-and-go.

d. Simulated flame-out landings on runways shorter than 6000 feet should be performed with extra caution.

e. The 160 knots approach speed for flame-out and simulated flame-out landings should be regarded as a minimum.

f. Slips should not be carried any closer to the ground than 600 feet.

g. Care should be taken to commence the flare-out at sufficient altitude to ensure a smooth landing.

h. When it appears that a hard landing is impending, the speed brakes should be retracted to prevent damage to the forward speed brake section.

After skill is once gained in the simulated flame-out landing, a pilot should refresh himself periodically to maintain a high level of proficiency in the maneuver. One or two simulated flame-out practices a month should be sufficient.

IMPORTANT

A simulated flame-out approach presents a good way to make an emergency landing when the reliability of the engine is questionable. Set up the standard simulated flame-out pattern. If at any time in the approach the engine stops, retract the speed brakes and continue the same approach pattern to a safe landing.

ENGINE FUEL CONTROL SWITCHOVERS.

It is suggested that the pilot make a few switchovers under static ground operation from primary to emergency fuel control at various throttle settings including OPEN to demonstrate that this safety feature is available.

Note

Return to the primary fuel control should always be done with the throttle at IDLE.

BELLY LANDING—ENGINE OPERATING.

If a normal landing cannot be made on a suitable airfield, a wheels-up landing in accordance with the following procedure is recommended:

a. Dump wing tank fuel. Jettison external stores, if carried.

Note

When airplane is decelerating, inertia forces will prevent fuel from being dumped from wing tanks.

b. Use up excess fuel to establish an aft cg condition and minimize possible fire hazard.

c. Move longitudinal control system selector switch to MANUAL.

d. Wing flaps—down. Reduce airspeed to approach speed.

e. Spread brakes—down. Extended speed brakes will cushion touch-down and minimize damage. Note that to extend speed brakes when the wing flaps are down, the speed brakes landing override switch must be ON.

f. Lock and tighten shoulder harness. Remove personal connections (oxygen, "G" suit, radio leads).

g. Open canopy.

h. Just before contact, retard throttle to CLOSED, set fuel master switch to OFF, and battery switch to OFF.

i. Make a normal landing, touching down as slowly as possible.

i. Abandon airplane immediately after it stops.

DITCHING—ENGINE OPERATING.

a. Dump wing tank fuel. Jettison external stores, if carried.

Note

When airplane is decelerating, inertia forces will prevent fuel from being dumped from wing tanks.

b. Use up excess fuel to establish an aft cg condition and to increase buoyancy in water.

c. Move longitudinal control system selector switch to MANUAL.

d. Landing gear control lever—UP.

e. Wing flaps—down. Reduce airspeed to approach speed.

f. Lock and tighten shoulder harness. Remove personal connections ("G" suit, radio leads).

g. Set oxygen regulator to 100% OXYGEN.

h. Open canopy.

i. Arresting hook may be extended, if desired, to give slight warning just before touch-down.

j. If sea conditions are favorable, land into the wind. If swells are running, land along a swell.

k. Just before contact, retard throttle to CLOSED. set fuel master switch to OFF, and battery switch to OFF.

l. Make a normal landing, touching down as slowly as possible in a nose-high attitude. Avoid high rates of descent in the final approach, since the airplane may not respond to flare-out immediately and may touch down hard.

m. Leave airplane as soon as motion definitely stops.

LANDING EMERGENCIES.

NOSE WHEEL UP OR UNLOCKED.

First, check for a hydraulic system emergency according to the procedures in Emergency Landing Gear Operation paragraph. After this, the flaps should be down and the speed brakes extended to minimize nose damage. If the main landing gear is down but the nose wheel is either up or down, but not locked, use up as much fuel as possible, drop external stores, if carried, and expend ammunition to get airplane to an aft cg, as light as possible and to reduce fire hazard.

a. Check that shoulder harness is tight and locked.

b. Disconnect oxygen, radio leads and "G" suit.

c. Open canopy (use emergency air bottle).

d. For carrier landings, extend arresting hook.

e. Accomplish a power-on, flaps-down approach, landing nose-high with as little sink speed as possible.

f. Move throttle to CLOSED just before touch-down.

Note

If the nose wheel is down but not locked, it will sometimes lock on landing. Caution should be used not to drop in hard enough to rotate onto the nose gear.

g. Set fuel master switch to OFF.

h. If landing area contains no obstructions and landing appears normal, do not set battery switch to OFF until airplane comes to a stop.

i. Hold nose off until speed is reduced, but let nose down onto runway before elevator effectiveness is lost.

j. Use brakes only if absolutely necessary.

k. Abandon aircraft as soon as it stops.

MAIN GEAR UP OR UNLOCKED.

First, check for a hydraulic system emergency according to the procedures in Emergency Landing Gear Operation paragraph. After this, the flaps should be down and the speed brakes extended to minimize nose damage. If one or both main wheels are not down and locked, use up as much fuel as possible, drop external stores, if carried, and expend ammunition to get the airplane as light as possible and to minimize fire hazard.

Note

Skidding of the airplane, rocking the wings and applying "g's" should be attempted as a means of locking down the main gear.

a. Check that shoulder harness is tight and locked.

b. Disconnect oxygen, radio leads and "G" suit.

c. Open canopy (use emergency air bottle).

d. For carrier landings, extend arresting hook.

e. Accomplish a power-on, flaps-down approach, landing nose-high with as little sink speed as possible.

f. Move throttle to CLOSED just before touch-down.

g. Set fuel master switch to OFF.

h. Set battery switch to OFF.

i. Hold airplane off as long as possible and, if one landing gear is down, land on side of runway next to extended landing gear.

Note

If one main landing gear is down but not locked, it will sometimes be locked if the landing is made in a skid towards the direction that will force the questionable landing gear outboard.

j. Abandon aircraft as soon as it stops.

BARRICADE ENGAGEMENT.

If all the arresting wires have been missed and a barricade engagement is imminent:

a. Keep the nose wheel on the deck.

b. Keep the airplane straight with the deck in order to contact the barricade nose first.

c. If time permits, shut down engine and set fuel master switch to OFF.

d. Keep head down and forward.

WARNING

Under unusual wing retardation barricade engagement conditions, such as free flight or severely yawed airplane attitudes, it is possible for the upper loading strap to enter the open cabin. IT IS THEREFORE MANDATORY FOR PILOT SAFETY THAT THE PILOTS KEEP THEIR HEADS DOWN AND FORWARD IN THE CABIN WHEN BARRIER AND/OR BARRICADE ENGAGEMENT IS IMMINENT. Such action can aid substantially in keeping the pilot's head and shoulders away from the back of the seat and headrest where a barricade strap is most liable to lodge if it should enter the cabin.

AIRSTART—J48-P-6A and J48-P-8 ENGINES.

NORMAL RELIGHTING IN THE AIR.

Note

All time delays mentioned in this section should be timed by clock.

If a definite flame-out has occurred, immediate relighting action must be carried out as follows:

a. Retard throttle to CLOSED.

Note

Due to the possibility of flooding the engine with fuel and of fuel fouling the spark igniters, which might prevent the igniters from sparking, this action should be carried out immediately to shut off the fuel flowing into the combustion chambers. However, do not close the fuel master switch, as this may cause the fuel system to become airlocked when the engine is windmilling and prevent a start.

b. Check that battery switch is at BAT. & GEN. position.

Note

When generator cuts out during flight (below about 35% rpm), all circuits except those necessary for flight will be disconnected automatically. Pitot head heat, communications circuits, etc, may be energized by setting the battery switch at BAT. Airstarts may be made with the battery switch in either BAT. & GEN. or BAT. position; BAT. & GEN. position is preferred to conserve power.

WARNING

Never set the battery switch to OFF under these conditions, and, when changing the battery switch setting from BAT. to BAT. & GEN., or vice versa, move it through OFF position as rapidy as possible if it has been necessary to set the engine fuel system selector switch to START & EMER. (J48-P-6A engines) or EMER. (J48-P-8 engines). Since the fuel control valve is held in the emergency system position electrically, setting the battery switch to OFF will release the holding solenoid and permit the control to return to the primary system, possibly causing another flame-out.

c. Set engine fuel system selector switch to START & EMER. position (J48-P-6A engines) or START & PRIMARY position (J48-P-8 engines).

Note

Using J48-P-8 engines, airstarts may be successfully performed with the engine fuel system selector switch in EMER. in cases where the primary fuel system is known to be malfunctioning or where all attempts to start the engine with the engine fuel system selector switch in the START & PRIMARY position have failed. The starting procedure, windmilling speed, and other conditions are the same for starts in either the START & PRIMARY or EMER. positions.

d. Cycle engine start master switch; i.e. set start master switch to OFF in order to reset ignition timer before each relight attempt, but be sure to set it to ON prior to relight attempt.

e. If above 25000 feet, use high altitude, high rpm airstart technique described in High Altitude, High RPM airstart paragraph, or reduce altitude to 25000 feet or below as fast as desirable, from a glide standpoint, to minimize engine cooling time. Approach 25000 feet at final glide speed of approximately 225 knots IAS. If below 25000 feet at time of flame-out, attain a speed of approximately 225 knots IAS by maintaining level flight or by gliding as necessary.

CAUTION

At least 30 seconds should be allowed prior to starting relight procedure to permit proper clearing of the engine and spark igniters of residual fuel. Failure to properly clear the engine is the primary cause of unsuccessful relight attempts.

f. When the altitude of 25000 feet or below, and the glide speed of 225 knots IAS have been attained, pull the nose up to maintain level flight, allowing the speed to decrease. When the airspeed is 195 KNOTS IAS, move throttle to the left of the CLOSED position to actuate the spark igniter system and hold at CLOSED position for 10 seconds.

g. Continuing to hold the airplane in level flight with speed decreasing, move throttle to IDLE position as soon as speed is reduced below 190 KNOTS IAS and the 10 second ignition delay is completed. This step should be completed in a range between 160-190 KNOTS IAS.

Note

The engine rpm will be a function of the airspeed. Light-up should take place at 175 knots IAS or below.

h. Holding altitude as constant as possible and the aircraft speed within the 160-190 knots IAS range, allow 30 seconds for light-up. Observe tailpipe temperature gage and tachometer for indications of light-up.

i. When light-up is obtained, keep throttle in IDLE position until engine has accelerated to altitude idle speed, then set throttle to desired engine speed.

Note

Using J48-P-6A engines, if primary fuel control operation is desired, set the engine fuel system selector switch momentarily to RESET PRIMARY, then to PRIMARY position prior to advancing the throttle. However, if it is believed that the flame-out was the result of primary fuel system malfunctioning, leave the fuel control in emergency operation.

j. If no light-up is obtained in 30 second period, or if engine does not continue to accelerate to idle rpm ("hung" start) in a reasonable time, RETARD THROTTLE TO CLOSED POSITION to shut off fuel to engine and again cycle engine start master switch to reset electrical timer. Increase airspeed to approximately 225 knots IAS by reducing altitude and repeat the level out and starting procedure outlined in steps f. through i. above. This maneuver will take 3000 to 5000 feet altitude.

Note

SUFFICIENT TIME MUST BE ALLOWED BETWEEN ELECTRICAL IGNITION RESTART ATTEMPTS TO PERMIT CLEARING OF THE ENGINE. A 30 second wait is considered mandatory for this purpose, and where conditions permit, 60 seconds is recommended.

EMERGENCY IGNITER OPERATING PROCEDURE.

Note

All time delays mentioned in this section should be timed by clock.

All engines are equipped with two single-shot, shell type, manually operated, emergency igniter units as an additional safety device for use in any case of failure to relight with the normal air starting ignition system. It is important to note that only two ignition attempts are available with this emergency ignition system. The cabin control is rigged so that the units may be fired either singly or in rapid succession, if desired.

Therefore, these emergency igniters should be used only when attempts to ignite with the standard air starting ignition system have failed or when time permits only one or, at the most, two possible relighting attempts. Conditions should be as identical as possible to those listed in the following procedure before starts are attempted with the emergency igniters. All starting attempts should be made with a single igniter except under extreme circumstances when time will permit only one attempt, in which case both igniters should be actuated.

The following procedure should be followed for all emergency igniter relight attempts:

a. Altitude—less than 18000 feet. Do not attempt emergency igniter relight at any higher altitude.

Airspeed—160-190 knots IAS.

Engine fuel system selector switch—START & EMER. (J48-P-6A engines) or EMER. (J48-P-8 engines).

b. Bring throttle from CLOSED to IDLE position.

c. Allow five second delay, then pull emergency igniter control.

d. Leave throttle in IDLE position until engine has reached altitude idle speed.

It should be noted that emergency igniter starts may be accomplished with J48-P-6A engines using either the START & EMER. or PRIMARY engine fuel system selector position, or with J48-P-8 engines using either the START & PRIMARY or EMER. position. However, in view of the limited number of emergency igniter starting attempts available (two shells), starting in the START & EMER. (J48-P-6A engines) or the EMER. (J48-P-8 engines) engine fuel system selector position is recommended unless the pilot suspects malfunctioning of the emergency fuel system.

Note

The procedures used with the emergency igniter system vary from those with the electrical ignition system in some details. Particularly to be noted is that with the emergency igniter system, the throttle should be placed in the IDLE position and a delay of five seconds allowed before discharging the igniter to ensure time for fuel to flow into the combustion chambers. With the electrical ignition system, the ignition switch is actuated and a delay of 10 seconds is made before the throttle is placed in the IDLE position to allow time for the spark igniters to warm up.

The "nose-down" procedure to attain increased speed, then "nose-up" for decrease of speed through the starting range, is not necessary in using the emergency igniter.

HIGH ALTITUDE, HIGH RPM AIRSTART.

Relights have been accomplished using the procedure outlined below, up to altitudes of 38000 feet. As soon as a flame-out is positively identified, initiate the following procdure:

a. Immediately return throttle to CLOSED and push outboard to START.

b. Then, immediately push throttle forward into IDLE detent. Relight will occur within two to four seconds.

c. If relight does not occur in eight seconds, return throttle to CLOSED and descend to 25000 feet where a normal airstart procedure can be initiated. If the primary fuel system is believed to have failed, causing the flame-out, move the engine fuel system selector switch to EMER. simultaneously with the actions outlined in step a. At 35000 feet and below, the procedure must be accomplished before the rpm drops below 40%. Above 35000 feet a progressively higher rpm is necessary to obtain a relight and approximately 75% rpm is necessary at 38000 feet. The airstart procedure must be performed in about four seconds at 38000 feet to obtain a successful airstart. It is absolutely necessary, however, to move the throttle into the START detent before moving it to IDLE. The pilot must have the procedure well in mind in order to initiate this airstart procedure immediately after the flame-out occurs.

More extensive information on high altitude, high rpm airstarts will be furnished when available.

FIRE.

ENGINE FIRE ON GROUND.

If one or both of the fire warning lights glow, indicating a fire in the engine compartment, or if there is other positive indication of fire:

a. Retard throttle to CLOSED.

b. Set fuel master switch to OFF.

c. After five seconds, set battery switch to OFF.

d. Leave cabin.

e. Use available fire extinguishers.

If a fire is detected in tailpipe after a normal shut down:

a. Call for external power.

b. Have ground crew observe fire for spreading (at a safe distance from end of tailpipe).

c. Perform a "clear engine", using procedure given in Clearing Engine paragraph in Section II.

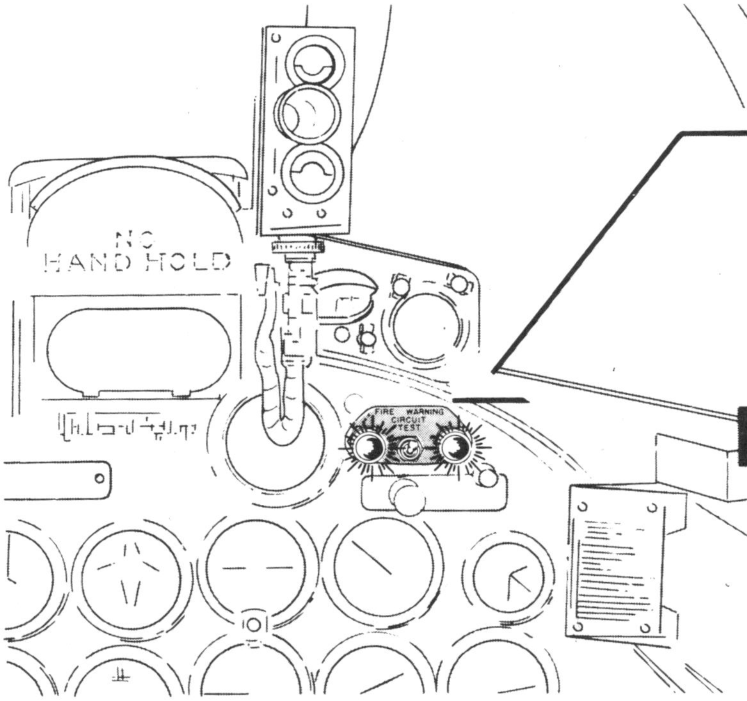

'FIRE'

Indications:

1. **Fire warning lights.**
2. **Smoke or fumes in cabin.**
3. **Visual indications of smoke or flame around the airplane or smoke trail behind the airplane.**
4. **Unusual engine roughness or muffled explosions.**
5. **Excess tailpipe temperature.**
6. **Information from wingman.**

Procedure:

1. **Retard throttle to CLOSED.**
2. **Set fuel master switch to OFF.**
3. **Make dead engine landing or abandon airplane.**

Figure 3-4. Fire

ENGINE FIRE IN FLIGHT.

If a fire should start in the engine compartment or aft sections of the airplane, the pilot will receive indications of it in at least one of the following ways:

a. One or both of the fire warning indicator lights will glow. Each light is connected to a separate circuit of fire detectors located at various points around the engine. When a fire warning light glows, it indicates that one of these detectors has been actuated. If there is reason to believe that the fire warning system is giving an erroneous indication, retarding the throttle to IDLE may cause the light(s) to go out. If this situation should arise, advance the throttle cautiously until the minimum speed required for flight is attained, then continue to the nearest base.

b. Smoke or fumes in the cabin.

c. Visual indications of smoke or flame around the airplane or smoke trail behind the airplane.

d. Unusual engine roughness or muffled explosions.

e. Exessive tailpipe temperatures.

f. Flickering arresting hook position warning light.

g. Information from wingman.

Usually a fire will make its presence known with more than one of the above indications. If only one indication appears, such as one fire warning light, fumes, smoke, or engine roughness, the trouble may not necessarily be a fire. Try to get at least two good indications of fire before taking emergency action.

If a fire is present, proceed as follows:

a. Retard throttle to CLOSED.

b. Set fuel master switch to OFF.

c. Make dead engine landing or abandon airplane.

ELECTRICAL FIRE.

The probability of electrical fires has been minimized by the provision of circuit breakers and fuses to protect the electrical circuits. However, if an electrical fire does occur, try to identify the source of the fire and pull the appropriate circuit breaker if it has not popped of its own accord. If the fire persists, set the battery switch to OFF and land as soon as possible.

SMOKE OR FUME ELIMINATION.

If smoke or any toxic fumes are present, proceed as follows:

a. Set oxygen regulator diluter valve lever to 100% OXYGEN.

b. Pull cabin pressure dump valve lever inboard. Operate cabin air conditioning system to obtain full ram air by setting system on-off switch to OFF (RAM AIR) and holding increase-decrease switch for at least six seconds. (See Cabin Air Conditioning System paragraph, Section IV.)

c. Remain on 100% oxygen until fumes or smoke have dissipated.

EMERGENCY ESCAPE.

Emergency escape from this airplane should be made by using the ejection seat with the following procedure:

a. REDUCE AIRSPEED IF POSSIBLE.

b. ACTUATE PRE-EJECTION LEVER. Actuate pre-ejection lever inboard, push it hard down and then push it outboard to lock. On early airplanes,[1] this jettisons canopy, lowers seat, releases knee braces, and pulls safety pin in seat catapult firing mechanism. On later airplanes,[2] this jettisons canopy, releases knee braces, and pulls safety pin in seat catapult firing mechanism, but does not lower the seat. (The radio/oxygen tube is pulled off the console nipple fitting to which it is attached and the radio lead severed by the force of ejection when the seat has traveled upward a distance of about 12 inches. The "G" suit hose connection is also separated during the initial upward travel of the seat.)

WARNING

Clearance between canopy and pilot's head is critical. Make sure head is below normal operating path of canopy before actuating pre-ejection lever.

On early airplanes[1], the ejection seat cannot be fired until the pre-ejection lever is actuated and the canopy is jettisoned.

On later airplanes[2], an emergency ejection seat arming control is provided. Under certain extreme flight conditions, such as high negative "g", it may be difficult to reach and actuate the pre-ejection lever. For this condition or one where the canopy fails to jettison, using the emergency ejection seat arming control will permit ejection through the canopy. The control may also be used to remove the safety pin in the event that the pin is not removed when the canopy is jettisoned. This control is a red handle on the left side of the headrest and is attached to the cable which normally pulls the safety pin in the seat catapult firing mechanism. Pulling the control forward past the pilot's left ear pulls out the safety pin and arms the seat. It does not lower the seat, nor does it release the knee braces. For this type of ejection, hold knees together to prevent injury.

WARNING

Ejection through the canopy is feasible only with the canopy in the fully closed position. The canopy cannot be closed if the pre-ejection lever has been actuated.

c. PLACE FEET ON FOOTRESTS.

Note

On early airplanes[1], the seat is lowered when pre-ejection lever is pulled. In this case, pilot slides his feet back and onto the footrests.

On later airplanes[2], the seat does not lower, but remains in a fixed position when the pre-ejection lever is pulled. In this case, the pilot may have to lift his feet up onto the footrests if the seat is in an elevated position.

Disconnect oxygen and radio leads.

d. ACTIVATE EMERGENCY OXYGEN SUPPLY. This action necessary only at altitudes above 20000 feet.

e. PULL FACE CURTAIN. While continuing to control airplane as much as possible, reach up with one hand and locate face curtain; then, sitting erect, with head back against headrest and muscles tensed to absorb ejection forces, reach up with the other hand and pull face curtain UNTIL FULLY EXTENDED. This will automatically lock the inertia reel and fire the ejection seat catapult.

Note

On early airplanes[1], the face curtain is pulled by means of two nylon rope loops attached to the face curtain. On these airplanes, the pilot's headrest is permanently attached to the movable portion of the seat and the position of the handles with respect to the pilot's head remains constant and is independent of seat position.

On later airplanes[2], the face curtain is pulled by means of a red, rubber covered double handle attached to the face curtain. On these airplanes, the pilot's headrest is fixed and is not attached to the movable portion of the seat. The position of the handles with respect to the pilot's head varies somewhat with seat position.

f. SEPARATE FROM SEAT Immediately following ejection, release face curtain, open lap belt, and roll forward out of the seat.

[1] F9F-6 airplanes ser No. 126257 through 127399 and F9F-6P airplanes ser No. 127473 through 128295.

[2] F9F-6 airplanes ser No. 127400 and subsequent and F9F-6P airplanes ser No. 128296 and subsequent.

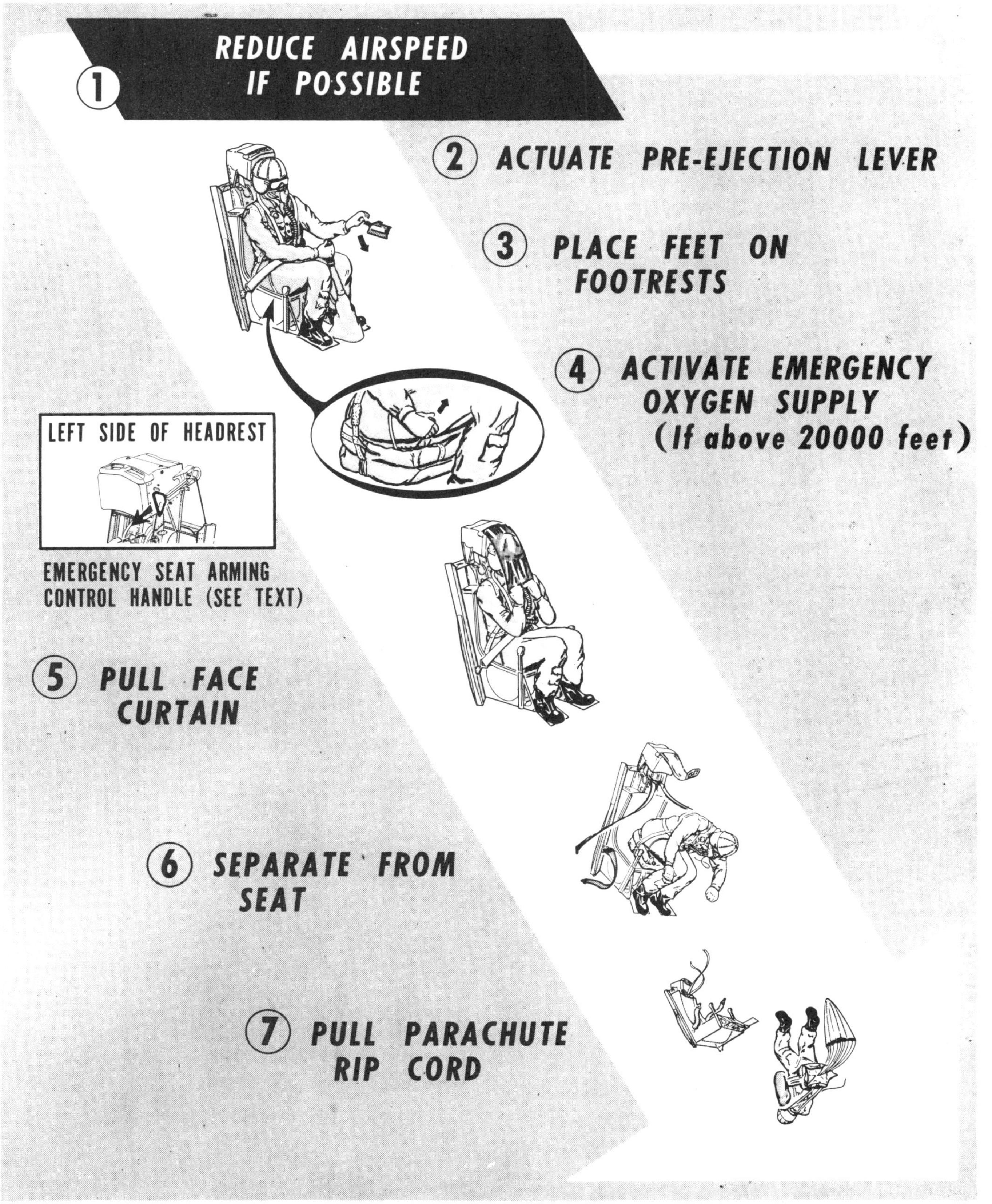

Figure 3-5. Emergency Escape

Note

If time permits, the oxygen tube and radio leads should be disconnected at the alligator clip coupling prior to ejection. When using the ejection seat for emergency escape at 2000 feet or below, it is recommended that the lap belt be opened and the oxygen tube and radio leads be disconnected prior to ejection. This procedure will shorten the time required for separation from the seat after ejection.

g. PULL PARACHUTE RIP CORD. If above 2000 feet at time of separation from seat, delay at least five seconds before pulling rip cord.

WARNING

This delay will preclude any possibility of collision between the seat and parachute and will also result in a lower parachute opening shock. When the escape is accomplished at high altitude, it is recommended that the pilot free fall to an altitude of approximately 18000 feet before opening his parachute in order to avoid the effects of anoxia and cold.

Do not pull rip cord while in seat.

Note

On airplanes with automatically lowering seats, the seat may be returned to its normal position after pulling the pre-ejection lever by returning the lever to its normal setting, at which time normal adjustment of the seat may be made. The firing mechanism will be armed after pulling the pre-ejection lever and care should be exercised not to pull the face curtain if ejection is not desired. The safety pin should be reinstalled after landing.

ENGINE EMERGENCIES.

PRIMARY FUEL CONTROL FAILURE.

Failure or malfunctioning of the engine primary fuel control may be indicated in several ways:

a. Flame-out—usually without warning and without excessive throttle movement. Amber emergency fuel system warning light on. Use relight procedure in emergency fuel control system.

b. Stuck or seized control—inability to change rpm.

c. Uncontrolled engine acceleration—rpm may increase slowly of its own accord from cruising rpm to 100% rpm or greater.

In cases b. and c., the emergency procedure is as follows:

a. Retard throttle to IDLE.

b. Set engine fuel system selector switch to START & EMER. (J48-P-6A engine) or EMER. (J48-P-8 engine).

c. Open throttle slowly to obtain desired engine speed.

d. Land as soon as practicable.

A flame-out may occur during switchover from the primary to the emergency fuel control systems, particularly at high altitudes, if there is a large difference between the rpm on primary at the time of switchover and the idle rpm on emergency. If a flame-out does occur, use standard relight procedures. In the case of uncontrolled acceleration, it is far better to risk the flame-out by switching systems than to risk damage to the engine by allowing it to run beyond its normal rpm and tailpipe temperature limits.

Failure or malfunctioning of one of the engine driven fuel pumps will be indicated by the engine fuel pumps warning light. If this light glows, and no other symptoms of trouble are present, remain in the primary fuel control system, throttle back to a conservative cruise rpm, and land as soon as practicable. If this light glows, and other symptoms of trouble are present, follow the procedure for switching to the emergency fuel control system presented above.

Note

It is preferable to remain on the primary fuel control system rather than to switch to the emergency fuel control system because the primary system incorporates automatic acceleration and decleration control. The emergency system consists simply of a metering valve in the fuel supply to the combustion chambers. When operating on the emergency fuel control system, both rpm and tailpipe temperatures must be watched carefully and kept within limits. Particular care must be used when changing altitude or when opening the throttle to increase power. High altitude operation is the most critical, as small throttle movements result in large changes in engine rpm.

Uncontrolled engine acceleration, commonly called "auto-acceleration", is primarily encountered at high altitudes and low temperatures when ice is formed while the fuel is flowing at a relatively slow rate through the servo system and being cooled by the compressor inlet air blast. The condition is normally corrected by a reduction of altitude or a changeover to the emergency fuel system.

AIRPLANE FUEL SYSTEM FAILURE.

If the low fuel boost pressure warning light glows, it indicates one of the following:

a. Both the submerged fuel boost pump in the forward fuel cell and the engine driven fuel boost pump have failed, or

b. The low pressure fuel filter has become clogged and is by-passing unfiltered fuel to the main pumps.

The recommended procedure is to remain on the primary fuel control system, throttle back to a conservative cruise rpm, and land as soon as practicable.

LOW FUEL QUANTITY.

With less than 400 pounds of fuel on board, special attention is required to maintain flight at low altitude. Attainable engine speed decreases as fuel is used, (see figure 1-6A). Fuel flow is approximately 80-130 pounds per minute under these conditions. The maximum attainable engine speed with 350 pounds of fuel is 92% rpm. The engine fuel pumps warning light will glow but should be ignored in this case. Obviously, the wave-off capabilities are decreased with less than 400 pounds of fuel indicated. It is recommended that a carrier landing be completed with at least 200 pounds of fuel remaining.

Note

The above paragraph applies to airplanes that do not have F9F Aircraft Service Change No. 286 incorporated. See Airplane Fuel System paragraph, Section I.

During an optimum descent (figure A-13), or minimum time descent (figure A-12) and any other rapid, idle rpm descent, fuel starvation will not occur if the descent is started with sufficient fuel for the descent plus landing. (Example: Optimum descent, 40000 feet to sea level, 180 pounds of fuel; one circle and landing, 250 pounds; minimum fuel on board at touch-down, 200 pounds. Total required at 40000 feet, 630 pounds.)

OIL SYSTEM MALFUNCTION.

If the oil pressure gage indicates abnormally low pressure, throttle back to lowest feasible rpm and land as soon as practicable.

If the oil pressure gage indicates zero, abnormally high pressure, or does not vary with rpm, replace the circuit fuse (right console, outboard of radio compass control panel). If trouble persists, follow procedure in preceding paragraph.

HYDRAULIC SYSTEM EMERGENCIES.

EMERGENCY HYDRAULIC SYSTEM OPERATION.

If there is insufficient pressure in the system to operate any of the units, set the selector valve control for the desired action and set the auxiliary hydraulic pump switch to ON. Hydraulic fuses are installed, which will isolate and make inoperable any system (i.e., landing gear, flaps, etc) in which a line break or leak occurs.

Note

It is possible that there may be a stoppage midway in the operation of one of the systems, caused by creeping of the fuse valves. To overcome this, set selector valve to reverse action for an instant, then return it to setting for action desired.

EMERGENCY FLAPERETTE OPERATION.

If the flaperons become inoperative due to insufficient hydraulic pressure, the switchover to the emergency flaperette system will be automatic. The only change the pilot will notice is a reduced rate of roll, since lateral control forces are the same for both the primary and emergency systems. The rate of roll on the emergency flaperette system is approximately one-third of that attainable on normal flaperon control at approach speeds in the landing condition and 10 to 20% of that attainable on normal flaperon control at moderate to high speeds. In the event that automatic changeover to the flaperette system does not occur, changeover may be made by placing the flaperette system switch to EMER. ON & TEST.

AIRPLANES EQUIPPED WITH ENGINE DRIVEN FLAPERETTE SYSTEM PUMPS.

Most airplanes are equipped with an engine driven flaperette hydraulic pump and an emergency flaperette air bottle system. During final approach to a dead engine forced landing, low windmilling engine rpm may not provide sufficient hydraulic pressure for unlimited use of the flaperettes for lateral control. To supplement output of the flaperette pump in this condition and also to provide a source of pressure for the rare case of a frozen engine, causing failure of both the flaperon and flaperette hydraulic systems, the emergency flaperette air bottle system is provided. Approximately twenty-two flaperette full deflection cycles are available when the system is turned on. To be sure that lateral control is available through the complete landing, use rudder for lateral control down to an altitude of 500 feet, depending on air turbulence, at which point actuate the flaperette air bottle system and use the flaperettes for lateral control in the normal manner.

AIRPLANES EQUIPPED WITH ELECTRICALLY DRIVEN FLAPERETTE SYSTEM PUMPS.

Certain early airplanes are equipped with an electrically driven flaperette hydraulic pump. No emergency flaperette air bottle system is installed in these airplanes. Regardless of whether the engine is windmilling or frozen, the electrically driven flaperette pump will provide adequate pressure for flaperette control through the complete descent and landing. With this system, it is important to conserve the battery as much as possible during an engine-off descent to be sure that electrical power is available to drive the flaperette pump when it is needed.

EMERGENCY LANDING GEAR OPERATION.

BY AUXILIARY HYDRAULIC PUMP.

If wheels fail to go down and lock when the normal control is moved to DOWN:

a. Check that airspeed is below limiting airspeed and that hydraulic combat system shut-off valve control is at HYDRAULIC PRESSURE ON (full pressure to landing gear system), and check hydraulic system pressure gage. Recycle gear (move control lever to UP, then return to DOWN) to check hydraulic fuses.

b. Set auxiliary hydraulic pump switch to ON and wait for an interval. Extension will be slow. Keep airspeed as low as practicable to increase extension speed.

BY EMERGENCY AIR BOTTLE.

If after attempting to lower the gear with the auxiliary hydraulic pump the indicator still does not show down and locked, complete speed brakes and flaps operations, reduce speed to 130 knots IAS maximum, then pull and lock emergency extension control handle (handle shaft is notched to lock handle in release position) to lower wheels by air pressure. The landing gear will lower in approximately five to six seconds.

WARNING

Effective on airplanes with F9F Aircraft Service Change 217 incorporated, pulling the emergency landing gear "T" handle will cause the normal hydraulic system to be dumped to the return line. Therefore, in a situation which will require landing gear emergency extension, efforts should be made to lower the flaps prior to extension of the landing gear.

Note

Once this system is used, the wheels will remain down and locked until pressure is relieved by a deck crew.

If the emergency extension control handle has been pulled before the flaps are lowered, it is possible to subsequently lower the flaps when the hydraulic trouble is limited to the landing gear sub-system. This action is not advisable because it requires returning the handle to its original position in order to close the landing gear dump valve. However, the operation can be performed as follows:

a. Return the handle to its original position.

WARNING

If the position indicator shows the landing gear to be unlocked after the emergency extension control handle has been returned to this forward (in) position, immediately pull the handle out again to pressurize the gear in the down position to prevent the gear from jamming the doors. If unlocking occurs, step b. below will not be possible.

b. Operation of flaps or other hydraulic services may then be completed.

c. After completing operations (step b.), pull and lock emergency extension control handle again to assure landing gear down and locked condition.

EMERGENCY SPEED BRAKE RETRACTION.

If speed brakes fail to return to up from an extended position:

a. Pull out speed brakes circuit breaker (located on left console, outboard of throttle quadrant).

b. Pull emergency speed brakes up control handle (red "T" handle on left side of seat bulkhead) and hold until brakes are up. If difficulty is due to an electrical failure in position selector system, retraction will occur in normal time of about five seconds. If difficulty is due to hydraulic failure, retraction time will be longer, since impact air pressure is being used to push the speed brakes up.

EMERGENCY BRAKE OPERATION.

The brakes are provided with a separate hydraulic reservoir and will operate after a pump failure or line failure elsewhere in the hydraulic system. The brakes are hydraulically boosted, however, and loss of hydraulic system pressure will make brake pedal forces about three times heavier.

The brakes will overheat and lose effectiveness if used excessively on landing or long taxiing. Effectiveness can be regained temporarily by pumping the brake pedal. More than six pumping strokes may be required under extreme conditions. To regain complete effectiveness, the brakes must be allowed to cool.

In the event of a brake system failure, an emergency air bottle system is provided. Pulling the red emergency "T" handle on the left console supplies air pressure to lock the brakes. (The "T" handle shaft is notched to lock handle in brakes on position.)

CAUTION

Since the air bottle pressure locks the brakes, do not use the emergency system if normal brakes will stop the airplane. If the air bottle is pulled while moving faster than 50 knots, blown tires will probably result. Braking effectiveness with blown tires is poor.

ELECTRICAL SYSTEM EMERGENCIES.

INDIVIDUAL CIRCUIT FAILURE.

If any electrical circuit fails to function, check that the appropriate circuit breaker reset button is in.

CAUTION

If button pops out repeatedly, do not attempt to hold it in, since continued use of the circuit may cause fire. Set the switch for that circuit off, if feasible.

GENERATOR FAILURE.

Generator failure will be indicated by the generator warning light and reduced voltage on the voltammeter. All circuits except those essential to flight (essential bus) will be automatically disconnected if the battery switch is in the BAT. & GEN. position and the landing gear control handle is in the UP position. Setting the battery switch to BAT., or setting the landing gear control handle to the DOWN position, will supply battery power to all circuits. Since the battery is the only source of power, it should be conserved by pulling all unnecessary circuit breakers when operating on battery power alone. (See figure 1-8 for detailed information.)

WARNING

Never set the battery switch to OFF except in case of an electrical fire.

TRIM FAILURES.

STABILIZER.

(See Longitudinal Control System Emergencies paragraph.)

WING OR RUDDER TRIMMER FAILURES.

There are no emergency procedures for failures in the wing trimmer or rudder trimmer systems, since they can be overpowered with normal flight controls.

Power for the wing trimmer passes through the stabilizer primary control circuit breaker on the right side of the seat bulkhead. If this circuit breaker pops or is pulled intentionally, the trim button on the control stick will become inoperative. In this case, set the stabilizer electrical power switch from NORM. to EMER. and use the stabilizer emergency control switch for stabilizer trim.

Power for the rudder trimmer passes through the flight instruments circuit breaker. If this circuit breaker pops or is pulled intentionally, certain flight instruments (landing gear position indicator, stabilizer trim position indicator, speed brakes position indicator, G-2 compass) will become inoperative.

LONGITUDINAL CONTROL SYSTEM EMERGENCIES.

Longitudinal control system emergencies may be caused by malfunction of either the hydraulic system of the flying tail or the electrical system of the adjustable (trimming) stabilizer. Therefore, it is of the utmost importance that the pilot determine which system is malfunctioning BEFORE taking any corrective action. In the event of a sudden longitudinal trim change when on hydraulic stabilizer, the pilot should move the stick correspondingly in order to determine what corrective action is necessary.

IF THE AIRPLANE RESPONDS TO STICK MOTION, IMMEDIATELY SET THE ELECTRICAL TRIM SELECTION SWITCH TO EMERG. AND EFFECT RECOVERY, USING THE EMERGENCY TRIM SWITCH. THE PILOT MAY CONTINUE THE FLIGHT IN HYDRAULIC LONGITUDINAL CONTROL UNTIL THE FLIGHT PLAN CALLS FOR A SHIFT, BUT DO NOT SHIFT TO MANUAL CONTROL UNTIL TRIM RECOVERY IS COMPLETE.

IF THE AIRPLANE DOES NOT RESPOND TO STICK MOTION, IMMEDIATELY SHIFT TO MANUAL CONTROL BY PUSHING DOWN ON THE EMERGENCY MANUAL CONTROL KNOB. THEN, IMMEDIATELY NOTE THE STABILIZER POSITION INDICATOR AND TRIM THE STABILIZER TO A NORMAL POSITION, USING THE CONTROL STICK TRIM SWITCH. REDUCE SPEED.

STABILIZER ACTUATOR RUNAWAY DUE TO ELECTRICAL FAILURE.

If an unusual change in longitudinal trim occurs while operating on flying tail but the airplane responds to stick motion (even though more than normal stick force is necessary), the uncalled for change in trim is due to stabilizer actuator electrical runaway.

DO NOT SHIFT LONGITUDINAL CONTROL FROM HYDRAULIC TO MANUAL. First attempt to regain proper trim by moving the electrical trim selector switch to EMER. and trimming by using the emergency trim switch. If movement of the emergency trim switch does not produce the desired trim change, set the selector switch to OFF. Fly at altitude to get the feel of the airplane at the speed of the landing approach and at touch-down.

The following is given as an example of the dangerous situation that will develop if the pilot misinterprets the cause of a stabilizer actuator runaway. Assume that the airplane is in level flight on the hydraulic longitudinal control system. A trim setting of one degree airplane nose-up is required to maintain level flight with neutral stick, zero stick force. Then an airplane nose-down electrical trim runaway occurs. To compensate for this nose-down trim change, the pilot moves the stick aft to maintain level flight and a one degree nose-up reading on the absolute stabilizer position indicator. The actuator runaway will continue until the actuator is at the maximum nose-down trim setting of 4-1/2 degrees. To maintain the one degree nose-up reading, the pilot will be holding the stick approximately eight inches aft of its neutral position. The stick force has become unusually heavy, which the pilot incorrectly interprets as a hydraulic malfunction and, therefore, decides to shift to manual elevator control. The stabilizer will immediately jump from one degree airplane nose-up to its trim setting of 4-1/2 degrees nose-down (a nose-down trim change of 5-1/2 degrees). The resultant violent pitch-over will probably cause the pilot to red-out and/or the airplane to undergo structural failure due to the negative "g" force. It would be impossible to pull out of the dive using manual elevator control alone.

The conditions described in the preceding example should have been corrected by the following procedure: The airplane responded to stick movement. Therefore, the pilot should have realized that the hydraulic system was functioning properly and that the nose-down trim change was caused by an electrical malfunction. Correct technique is to attempt to regain trim control by using the emergency trim circuit. If trim is not recovered, set the trim selection switch to OFF and follow the procedure described below.

WARNING

Never shift from hydraulic to manual control while holding the stick out of neutral, unless the airplane is NOT responding to stick movement.

If the stabilizer runs to and sticks at full airplane nose-down trim (1-3/4 degrees), a flaps-up approach for landing is preferred to avoid the nose-down trim change with the flaps down. Since an aft cg (light fuel loading) is most desirable, delay the landing, if possible, to use up fuel. Stick forces will be high (approximately 25 pounds pull). The landing should be made on the powered longitudinal control system, keeping the wing flaps retracted (to avoid control shift into manual). With the flaps retracted, the approach should be about 15-25 knots faster than with flaps down, i.e., maintain 140 knots IAS for field landings, 140 knots IAS for carrier landings. With 140 knots carrier approach airspeed, the wind over the deck must be high (see figure 6-7).

If the stabilizer runs to and sticks at full airplane nose-up trim (6 degrees), a flaps-down approach is preferred to take advantage of the nose-down trim change with the flaps down. Since a forward cg (heavy fuel loading) is most desirable, land without unnecessary delay. Stick forces will be moderately high (approximately 15 pounds push).

STABILIZER ACTUATOR RUNAWAY DUE TO HYDRAULIC FAILURE.

If an unusual change in longitudinal trim occurs while on flying tail but the airplane does NOT respond to stick motion, the uncalled-for change in trim is due to a stabilizer actuator hydraulic runaway. IMMEDIATELY SHIFT TO THE MANUAL ELEVATOR CONTROL SYSTEM by pushing down the emergency manual control knob. The shift from hydraulic control to manual occurs in approximately 1 to 1-1/2 seconds and automatically shuts off pressure to the flying tail components. The position of the stabilizer, as indicated by the absolute stabilizer position indicator, should be noted immediately and brought to normal by electrical trim. Speed should be reduced to decrease stick forces and increase effectiveness of manual control.

Note

Pushing the emergency manual control knob down not only shifts longitudinal control to manual elevator control but it also by-passes the 2-1/2 degrees airplane nose-up trim limit circuit and the automatic retrimming feature with flap retraction. This last fact should be remembered in the event of a wave-off.

If the shift to the manual control does not occur when the emergency manual control knob is pushed down, the pilot may, after speed is decreased to below 220 knots IAS, extend the flaps and thereby cause the shift to manual elevator control.

CROSS-TRIMMING.

Cross-trimming is a condition where the stabilizer is hydraulically held away from the 6 degree airplane nose-up or 1-3/4 degrees nose-down limits while it is trimmed towards them electrically (pulling aft on the stick while trimming nose-down, or vice versa). Cross-trimming may be caused by pilot inattention or confusion, and subsequent attempts to hydraulically compensate for opposite electrical trim. The trim limits can be exceeded and the trim can be run to the limit of actuator travel (6-1/2 degrees airplane nose-up or 4-1/2 degrees airplane nose-down).

Continued flight on hydraulic longitudinal control will be normal as long as stick force opposing trim is held. However, shifting from hydraulic to manual elevator control when in the cross-trimmed condition will cause a large semi-uncontrollable pitch, as the stabilizer moves instantaneously to its trim position.

Cross-trimming can also result from a stabilizer electrical actuator runaway (see Stabilizer Actuator Runaway Due to Electrical Failure paragraphs).

The following is given as an example of the dangerous situation that may develop if the pilot cross-trims. Assume that the airplane is starting a climb from level flight at 300 knots at 5000 feet while on hydraulic control. Upon applying back pressure to the stick, his gloved hand accidentally pushes the stick grip trim switch forward. The nose-down trim resulting from the forward push on the trim switch causes the pilot to pull back harder on the stick. Sensing something wrong, the pilot erroneously decides to shift to manual control and pushes down the emergency manual knob or sets the longitudinal control system selector switch to MANUAL. This causes the stabilizer to move instantly upward to its trim setting. The airplane may pitch down violently and cause the pilot to red-out.

The conditions described in the preceding example should have been corrected by the following procedure: Before pushing down the emergency manual knob or setting the longitudinal control system selector switch to MANUAL, the pilot should have determined what was malfunctioning. He should have pulled the stick back to see if the nose would rise, thus indicating that he had hydraulic control or, if he had full back stick, the airplane nose-down response to a push on the stick would have indicated that he had hydraulic control. He then should have set the electrical trim selection switch to EMERG. immediately and effected recovery using the emergency trim switch.

NORMAL ELECTRICAL TRIM SYSTEM FAILURE.

If the stabilizer electrical trim fails, as indicated by a

lack of response to movements of the control stick grip trim switch, move the electrical trim selector switch on the left console from NORM. to EMERG. Trim the stabilizer by using the emergency trim switch located aft of the selector switch.

TOTAL ELECTRICAL TRIM SYSTEM FAILURE.

No response to movement of either the stick grip trim switch or emergency trim switch indicates a total failure of the stabilizer electrical trim system. In the event of an uncalled-for trim change (which may be caused by fusing of the electrical wiring running to the actuator) that cannot be corrected by use of either the normal or emergency switch, set the electrical trim selector switch to OFF. This will stop any further actuator movement. In either situation, if at a safe altitude, determine the amount of out-of-trim and whether the airplane can be controlled in the landing configuration on manual elevator control. If control is unsatisfactory, the flaps-up landing should be made, using hydraulic control as described in the paragraphs under the heading Stabilizer Actuator Runaway Due to Electrical Failure.

SLUGGISH RESPONSE OR TOTAL LACK OF RESPONSE TO STICK MOVEMENT.

Sluggish response or total lack of response to stick movement indicates a hydraulic control system failure. The pilot should immediately shift to the manual elevator control system by pushing down the emergency manual control knob. The shift from hydraulic control to manual automatically shuts off pressure to the flying tail components. The position of the stabilizer, as indicated by the absolute stabilizer position indicator, should be noted immediately and brought to a normal position by electric trim. Speed should be reduced to decrease stick forces and increase effectiveness of manual control.

Note

Pushing down the emergency manual control knob not only shifts longitudinal control to manual elevator but it also by-passes the 2-1/2 degrees airplane nose-up trim limit circuit and the automatic retrimming feature with flap retraction. This last fact should be remembered in the event of a wave-off.

If the shift to manual control does not occur when the emergency manual control knob is pushed down, the pilot may, after speed is decreased to below 220 knots IAS, extend the flaps and thereby cause the shift to manual elevator control.

AUTOMATIC SHIFT TO MANUAL CONTROL.

In the event of complete loss of hydraulic system pressure, a safety feature is incorporated to automatically shift the longitudinal control from the hydraulically operated stabilizer to the manual elevator control system. The pilot will be made aware of this occurrence because of the increase in stick forces.

EMERGENCY ARMAMENT CONTROL—F9F-6 AIRPLANES.

a. To stop runaway guns, set armament master switch to OFF or inboard and outboard gun charging switches to SAFE.

b. To jettison bombs, rocket launchers or drop tanks from MK 51 racks, move stores jettison switch up, or pull emergency hydraulic jettison "T" handle aft.

EMERGENCY DROP TANK JETTISON CONTROL—F9F-6P AIRPLANES.

To jettison drop tanks from MK 51 racks, raise guard on drop tanks jettison switch on left console and move switch outboard, or push down emergency hydraulic stores release lever on left cabin rail.

EMERGENCY OPERATION OF IFF EQUIPMENT.

To indicate emergency or distress, press the dial stop (see figure 4-4) and rotate the master selector to its EMERGENCY position.

SECTION IV

TABLE OF CONTENTS

SECTION IV DESCRIPTION AND OPERATION OF AUXILIARY EQUIPMENT

CABIN AIR CONDITIONING SYSTEM.

Hot, high pressure air is provided by the engine compressor. Part of this air is passed through a refrigeration unit and cooled, then mixed with the remainder and fed to the cabin. Cabin temperature is regulated by the ratio of hot air to cool air. This regulation is accomplished by the use of a valve that controls the amount of engine air that by-passes the refrigeration unit. The amount of air fed into the cabin varies with engine rpm and altitude. A cabin pressure regulator maintains pressure control by venting the excess air. The cabin is pressurized only above 5000 feet (figure 4-2); however, temperature control is available on the ground as well as at any altitude. The cabin pressurization system has only the normal pressure schedule.

NORMAL CONTROLS.

SYSTEM CONTROL PANEL.

The air conditioning system control panel, located on the left console, contains the system on-off (temperature-ram air) and the increase-decrease switches.

SYSTEM ON-OFF SWITCH.

This switch permits the pilot to select conditioned air or ram ventilating air for the cabin. The switch may be set to either the ON (TEMP) position, which provides temperature control, or the OFF (RAM AIR) position, which provides ventilation. The main shut-off valve, through which hot engine bleed air flows into the air conditioning system, is positioned by this switch. When the switch is set to ON, the main shut-off valve (17, figure 4-1A) will open and the ram air shut-off valve (16, figure 4-1A) will close if it is open. The actuator for the main shut-off valve rotates in one direction only and is spring loaded to close. This drive mechanism, however, does not permit the valve to go closed unless it has reached a full open condition. Once the system on-off switch is placed in the ON (TEMP) position, the system cannot be turned off until the main shut-off valve has completed its opening cycle travel. The opening cycle requires 15 to 22 seconds and if the on-off switch is set to OFF prior to completion of this cycle, the valve will be stopped in a partially open position and the pressurization system will continue to operate. If the main shut-off valve is full open and the on-off switch is set to OFF (RAM AIR), the main shut-off valve will close and the by-pass valve (12, figure 4-1A) will open.

FIGURE 4-1 DELETED

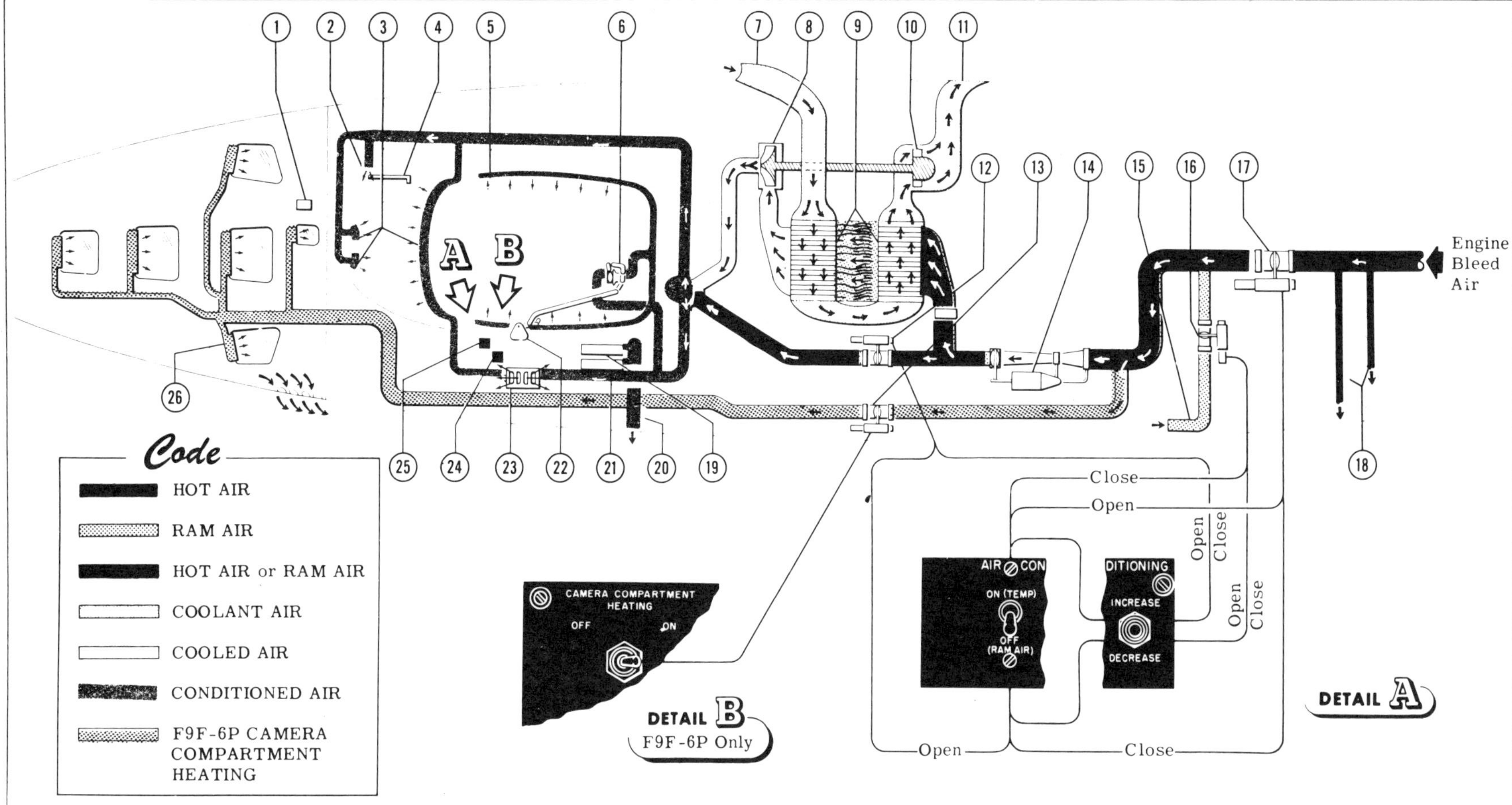

1. Camera Compartment Automatic Temperature Control (F9F-6P)
2. Foot Outlet
3. Windshield Defroster Outlets
4. Foot Outlet Control
5. Canopy Defroster Outlet
6. Canopy Defroster Shut-off Valve
7. Coolant Air Scoop
8. Coolant Fan Turbine
9. Refrigeration Unit
10. Coolant Fan
11. Coolant Air Exhaust
12. By-pass Valve
13. Camera Compartment Temperature Control Valve (F9F-6P)
14. Flow Control Valve
15. Ram Air Scoop
16. Ram Air Shut-off Valve
17. Main Shut-off Valve
18. Engine Bleed Air to Pressurize Hydraulic Tanks and Anti "G" Suit and Operate Turn and Bank Indicator
19. Cabin Pressure Manual Dump Valve
20. Pressure Regulator and Dump Valve Exhaust
21. Cabin Pressure Regulator Valve
22. Canopy Defroster Shut-off Valve Control
23. Console Outlet and Control
24. Console Camera Compartment Heat Control Panel (F9F-6P)
25. Console Air Conditioning System Control Panel
26. Camera Compartment Window Defroster Outlets (F9F-6P)

Figure 4-1A. Air Conditioning System Schematic Diagram

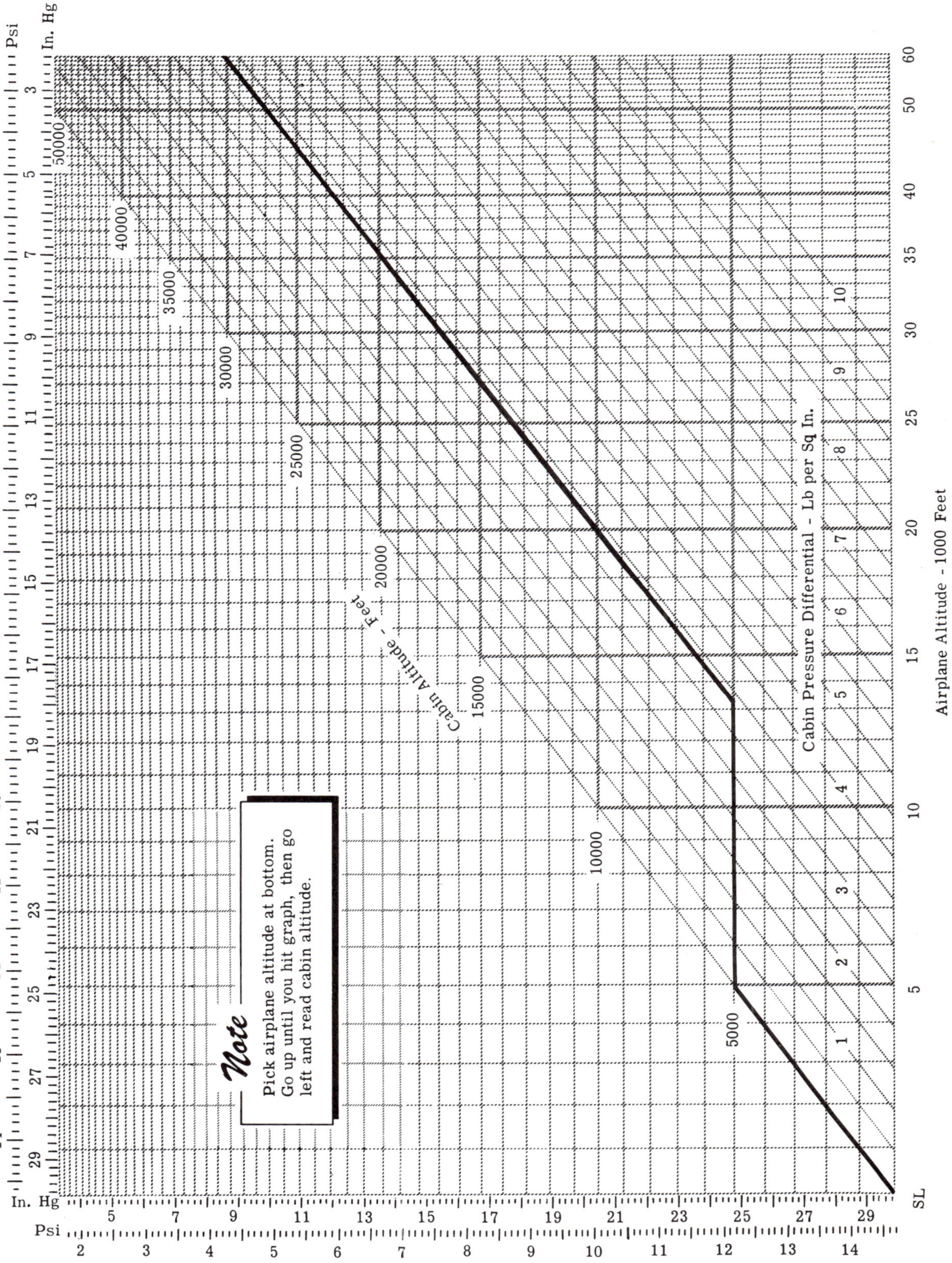

Figure 4-2. Pressurizing Schedule

INCREASE-DECREASE SWITCH.

When the on-off switch is in the ON (TEMP) position, the two position momentary (center off) increase-decrease switch permits the pilot to regulate the ratio of hot to cool air entering the cabin. This switch operates the by-pass valve. Full valve adjustment is made in 20 seconds. When the switch is held at INCREASE, the by-pass valve opens and allows more hot air to enter the cabin, thus increasing the cabin temperature. When the switch is held towards DECREASE, the valve closes and more air is forced through the refrigeration unit, thus decreasing the cabin temperature.

When the on-off switch is set in the OFF (RAM AIR) position, the increase-decrease switch permits the pilot to regulate the amount of ram ventilating air that enters the cabin. Holding the switch at INCREASE opens the ram air valve, and holding the switch towards DECREASE closes the valve, thus regulating the amount of ram air. The ram air valve will be closed when the on-off switch is first placed in the OFF (RAM AIR) position. Full valve adjustment is made in approximately four seconds.

MANUAL CABIN AIR OUTLET CONTROLS.

The controls consist of a sliding cover plate controlling an outlet on the right side forward of the console, at the pilot's feet, and a sliding shutter regulating a louvered outlet on the left console. The volume of ventilation air entering the cabin from these outlets may be regulated as desired by adjusting the respective control (14, figure 1-3 and 29, figure 1-3B).

CANOPY DEFROSTER SHUT-OFF VALVE CONTROL.

This control consists of a lever mounted on the left side of the canopy. It is connected through a mechanical linkage to a butterfly valve in the canopy defroster duct. Pushing the lever forward to OFF closes the valve and moving it aft to ON opens the valve (13, figure 1-3).

Note

For maximum windshield defogging and defrosting, particularly when descending from high altitude into warm humid air, manually adjust the cabin air to full hot and close the foot, console and canopy valve outlets before starting descent. This will force the maximum amount of hot air to the thick armor glass without excess heat to the pilot. After descent, readjust temperature as dictated by frost or pilot comfort.

EMERGENCY CONTROL.

CABIN PRESSURE DUMP VALVE.

The cabin pressure dump valve lever is mounted on the left side of the bulkhead behind the pilot's seat. Pulling the lever inboard instantly dumps the cabin pressure, and pushing it outboard restores pressure. Cabin temperature control is still available after the pressure has been dumped (7, figure 1-3).

INDICATOR.

CABIN PRESSURE ALTIMETER.

This instrument is mounted on the vertical panel extending from the forward end of the right console. It indicates the cabin pressure altitude and serves as a check for the proper operation of the cabin pressurization system (24, figure 1-3A and figure 4-2).

OPERATING INSTRUCTIONS.

a. Set on-off switch to ON (TEMP) if temperature control and cabin pressurization is desired, or to OFF (RAM AIR) if ram air ventilation only is desired.

b. Move momentary increase-decrease switch to position which will produce the state of air conditioning desired for optimum pilot comfort.

Note

Under certain atmospheric conditions of high temperature and humidity, chiefly at low altitudes, the cabin air conditioning system outlets will emit fog. This fog becomes thicker as the cabin temperature falls, but can be eliminated by placing the on-off switch to ON (TEMP) and moving the increase-decrease switch to INCREASE. The fog may be dense enough to resemble smoke; therefore, to avoid cause for undue alarm, until the pilot has become accustomed to the characteristics of the air conditioning system, take-offs should be made with the on-off switch set to OFF (RAM AIR) and the quantity of ventilating air required to clear the cabin controlled by the increase-decrease switch.

c. For ram air ventilation, set on-off switch to OFF (RAM AIR) and pull cabin pressure dump valve lever inboard. Regulate quantity of ventilating air by holding increase-decrease switch at INCREASE position until desired flow of ram air is obtained.

d. To dump pressure, PULL dump control handle (left of seat) inboard; to restore pressure, PUSH outboard.

EMERGENCY PRESSURIZING EQUIPMENT OPERATION.

In case of excessive pressure; PULL dump control handle, located to the left of the seat, inboard to DUMP pressure. The cabin air conditioning control will continue to function. In case of excessive cabin temperature: set on-off switch to OFF (RAM AIR) and move increase-decrease switch to INCREASE. When temperature has been reduced, attempt to regain desired cabin air condition. If ineffective, return to OFF (RAM AIR) position.

ELECTRONIC EQUIPMENT.

UPON ENTERING CABIN.

a. Plug headphones and microphone into pilot's quick disconnect assembly.

TABLE IV-I.
COMMUNICATION AND ASSOCIATED ELECTRONIC EQUIPMENT

Type	*Designation*	*Function*	*Range*	*Location Of Controls*
VHF Transmitter-Receiver or	AN/ARC-1	Two-way voice communication	Line of sight	Right console
UHF Transmitter-Receiver	AN/ARC-27, -27A	Two-way voice communication	Line of sight	Right console (See Note)
Homing Adapter	AN/ARA-25	Navigation or homing	Line of sight	Right console (See Note)
Radio Compass	AN/ARN-6	Navigation (ADF)	100 statute miles	Right console (Indicator on instrument panel)
IFF Equipment	AN/APX-6	Identification	Line of sight	Right console
Receiver	AN/ARR-2A	Navigation—homing	Line of sight	Right console
Radio Altimeter[1]	AN/APN-1	Indicator—altitude above terrain and water	4000 feet	Left console (Indicator on instrument panel)
Radio Altimeter[2]	AN/APN-22	Indicator—altitude above terrain and water	10000 feet, land 20000 feet, water	Left console (Indicator on instrument panel)
Gun Ranging	AN/APG-30	Range indication for sight unit	3000 yards max	Left console

Note

The AN/ARA-25 system utilizes AN/ARC-27, -27A controls and the course indicator on the instrument panel.

b. Turn battery and radio master switches ON. The radio master switch is in the C-739/ARC master control unit on the radio console. The radio master switch does not supply power to the IFF equipment, nor to the radio altimeter equipment (F9F-6P).

Note

The C-739/ARC is not installed when the AN/ARC-27 is installed.

c. Set COMM VOLUME control to maximum (full clockwise).

AN/ARC-1—VHF TRANSMITTER-RECEIVER.

OPERATION—RECEIVER.

For an operating check of this equipment, it is necessary that signals be present on the channels on which operation is contemplated. In the absence of signals, the squelch circuit reduces the receiver output to zero, and therefore, it is impractical to attempt to determine proper receiver performance.

On the vhf console control unit C-865/ARC-1, operate the rotary selector switches as follows:

a. For reception on any of the nine main channels, use position MAIN T/R and rotate CHAN SEL switch to desired channel.

b. For reception on any one of the nine main channels and the guard channel, use position BOTH and rotate CHAN SEL switch to desired channel.

c. For reception on guard channel only, use position GUARD. Position of CHAN SEL switch need not be considered.

The volume level of the vhf receiver is controlled by the COMM VOLUME on the master control unit. At the completion of this check, set the volume to minimum.

OPERATION—TRANSMITTER.

Note

These instructions are subject to local limitations regarding radio silence.

a. On vhf console control unit, operate rotary selector switches as follows: for transmission on any one of the nine main channels, use positon MAIN T/R and rotate CHAN SEL to desired channel.

b. For transmission on guard channel, use position GUARD. Position of CHAN SEL switch need not be considered.

c. For transmission and reception on any main channel and simultaneous reception on guard channel, use position BOTH.

d. To transmit, press throttle microphone switch and talk.

e. Release microphone switch to receive. If it is not desired to monitor radio compass or navigation receiver while transmitting, reduce associated SENS control.

AN/ARC-27, -27A UHF COMMAND SYSTEM.

DESCRIPTION.

The AN/ARC-27, -27A command equipment provides AM radio communications between the airplane and any other station similarly equipped within the ultra high frequency range of 225.0 to 399.9 megacycles. This system provides 1750 channels, from which any 18 or 20 channels, depending on the control equipment installed in the airplane, may be made available to the

[1]F9F-6P airplanes ser No. 127473 through 127488.
[2]F9F-6P airplanes ser No. 127489 and subsequent.

pilot. In addition, a guard channel frequency is available. The system consists of an RT-178/ARC-27 transmitter-receiver, an antenna and associated wiring, and one of three control systems. On early airplanes,[1] a C-628/ARC-27 console control panel and a C-626/ARC-27 master (channel setting) control box are installed. On later airplanes,[2] a C-905/ARC-27 console control panel is used. The final series of airplanes are equipped with a C-1015/ARC-27A console control panel and the C-626/ARC-27 is not used.[3]

This equipment may be operated as a navigational aid when used in conjunction with AN/ARA-25 equipment (see AN/ARA-25 paragraph).

CONTROLS.

The C-628/ARC-27[1] console control panel provides the controls for operating the AN/ARC-27 uhf command system. These controls are the OFF-T/R-T/R+G REC.-ADF selector switch which turns the set on for reception and transmission with or without guard channel reception, a volume control and a channel selector switch. The C-905/ARC-27[2] console control panel provides a MANUAL-PRESET switch and three decade frequency dials marked MANUAL CHANNEL, in addition to three controls identical to the three controls on the C-628/ARC-27 control panel. With the MANUAL-PRESET switch set at PRESET, any one of nineteen preset channels may be selected for reception and transmission by means of the channel selector switch marked PRESET CHANNEL. With the MANUAL-PRESET switch set at MANUAL, the PRESET CHANNEL control is disabled and selection of any one of the 1750 uhf channels may be accomplished by means of the three decade frequency dials marked MANUAL CHANNEL.

Note

Console control panels C-628/ARC-27 or C-905/ARC-27 are also used to control the AN/ARA-25 homing adapter if the latter is installed.

The C-1015/ARC-27A[3] console panel provides the controls for operating the AN/ARC-27A uhf command system and the AN/ARA-25 homing adapter. These controls are: the OFF—T/R—T/R+G—ADF selector switch, a VOL control, a SENS control, a CHAN selector switch, three rotary decade frequency selector dials, and a mechanical push button.

The OFF-T/R-T/R+G-ADF selector switch turns on the equipment when advanced from the OFF position. When set to T/R, only the transmitter-receiver operates, and when set to T/R+G, the transmitter-receiver operates together with the guard receiver. In the ADF position, the uhf homing adapter is put into operation. When set to ADF, the transmitter-receiver is still in an operational condition on whatever channel or frequency is set by the CHAN selector switch or, if this switch is in manual position, the rotary decade frequency selector dials.

The VOL control is used to adjust the output of the receivers and the SENS control is used to reduce audio background noise to a minimum.

The CHAN selector switch is used to select any one of 20 preset channels and one preset guard channel. When the switch is set to M, the three rotary decade frequency selector dials may be used to select any desired frequency without danger of changing the preset channels.

The mechanical push button marked PUSH TO SET CHAN is used in conjunction with the rotary decade frequency selector dials to preset any desired frequency on any desired setting of the channel selector switch (see Channel Preset Procedure paragraph).

CAUTION

Do not press the mechanical push button in unless setting up a specific frequency. The switch is mechanical and does not require that the equipment be on to change preset settings.

CHANNEL PRESET PROCEDURE.[3]

a. Set CHAN selector switch to 1.

b. Rotate the three rotary decade frequency selector dials to desired frequency (indicated in line with frequency selector pointer).

c. Turn mechanical push button clockwise to its stop, push full in and release.

d. Repeat preceding steps for setting each desired channel. G position on channel selector switch is used for guard channel.

Note

Do not set up any frequency outside the specified range of the equipment.

e. If possible, check setting by establishing two way communications on new frequencies.

CAUTION

Do not press the mechanical push button in unless setting up a specific frequency. The switch is mechanical and does not require that the equipment be on to change preset settings.

OPERATION.

REMOTE RECEPTION AND TRANSMISSION.

Note

Operation of the AN/ARC-27 equipment using any one of the three control panels which may be installed is as described below.

a. Set battery switch to BAT. & GEN.

[1]F9F-6 airplaes ser No. 126257 through 127225.

[2]F9F-6 airplanes ser No. 127226 through 127326, 127359 through 128216 and F9F-6P airplanes ser No. 127473 through 128310.

[3]F9F-6 airplanes ser No. 127327 through 127358, 128217 and subsequent, and F9F-6P airplanes ser No. 131252 and subsequent.

b. Set uhf console control panel switch to channel desired and T/R. In approximately one minute, signals above the squelch level will be heard.

c. To hear signals on both guard and channel selected, set uhf console control panel selector switch to T/R+G REC.

d. To transmit on channel selected, depress microphone switch and talk into microphone. Sidetone will be heard in headset.

e. To transmit on guard channel, set selector switch to G and repeat steps above.

STOPPING EQUIPMENT.

On the uhf console control panel, set the OFF-T/R-T/R+G-ADF switch to OFF.

Note

> Either AN/ARC-1 or AN/ARC-27 equipment may be installed, but not both.

AN/ARA-25 UHF HOMING ADAPTER.

This system is used in cojunction with the AN/ARC-27 command set to serve as a homing or navigational aid by providing a continuous indication of the relative bearing of an r-f signal source transmitted by another airplane, surface craft, or ground station on a frequency range of 225.0 to 399.9 megacycles. The relative bearing is indicated by the number 1 pointer on the course indicator mounted on the instrument panel.

Whenever the OFF-T/R-T/R+G-ADF selector switch on the AN/ARC-27 control panel is set to the T/R or T/R+G position, the AN/ARA-25 equipment is automatically put in the stand-by condition. When this selector switch is set to ADF, the AN/ARC-27 equipment is placed in circuit with the homing equipment, enabling the latter to receive the frequency selected by the CHAN selector switch on the AN/ARC-27 control panel.

Note

> When the OFF-T/R-T/R+G-ADF selector switch is set to ADF, AN/ARN-6 radio compass information is indicated by the number 2 pointer on the course indicator, while AN/ARA-25 information is indicated by the number 1 pointer. When this switch is set to either T/R or T/R+G, or when the microphone switch is depressed to transmit, uhf homing information is switched off automatically, and both needles join to display radio compass information on the course indicator.

OPERATION.

a. Set OFF-T/R-T/R+G-ADF selector switch on AN/ARC-27 control panel to ADF. After three minutes, background noise will be heard in headset indicating that equipment is operating.

b. Set CHAN selector switch on AN/ARC-27 control panel to frequency of known station.

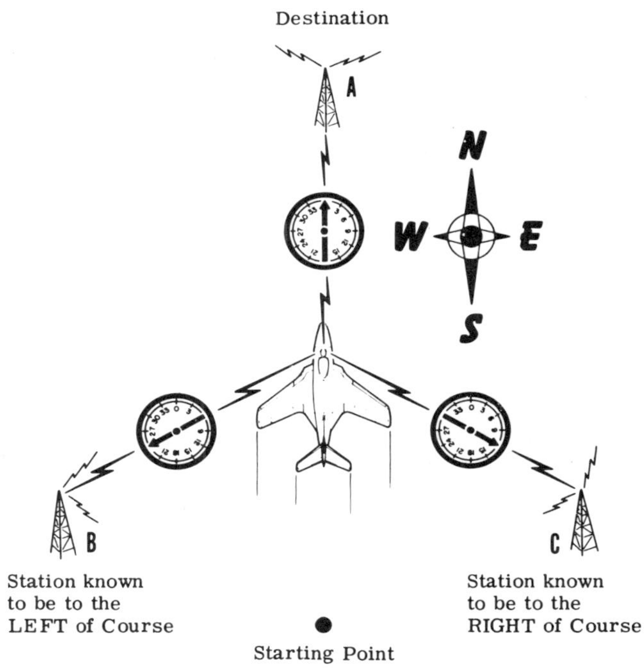

Fix point located on chart by projecting lines from stations at angle with true north shown by indicator.

Figure 4-3. Radio Compass Position Diagram

c. Number 1 pointer on course indicator will give relative bearing of station and audio from station will be heard in headset.

d. Homing equipment will revert to stand-by condition when microphone switch is depressed to transmit or when OFF-T/R-T/R+G-ADF selector is moved to T/R or T/R+G.

e. To turn off equipment, set OFF-T/R-T/R+G-ADF switch to OFF.

AN/ARN-6—RADIO COMPASS.

STARTING.

Turn function switch on C-149/ARN-6 (on right console) to COMP., ANT. or LOOP position.

STOPPING.

Turn function switch to OFF.

OPERATION.

GENERAL.

The equipment will perform the following three major functions:

a. Homing compass operation.

b. Position finding, using automatic and aural-null methods.

c. Receiver operation, using ANT. or LOOP.

HOMING COMPASS OPERATION.

To use as a homing compass, perform the following operations:

a. Turn function switch to COMP.

b. Rotate band switch to frequency band in which operation is desired.

c. Turn tuning crank to desired station frequency and tune for maximum swing of tuning meter. Greater accuracy in tuning may be obtained by setting CW-VOICE switch to CW. A 900 cycle tone will be heard along with the station modulation. This will aid in accurate tuning. After tuning, return CW-VOICE switch to VOICE to eliminate the 900 cycle tone.

d. Adjust AUDIO control for desired headset level.

e. Listen for station identification to be sure that correct station is being received.

f. Turn VAR knob on indicator until the azimuth zero is at the index.

g. The indicator pointer will now show the bearing of the station relative to the airplane heading. For example, if the pointer is to the left of zero, the station is on your left. Turn the airplane to the left until the pointer is at zero. If the airplane heading is held at zero degrees on the radio compass indicator, you will ultimately fly over the radio station antenna. Cross winds, however, will cause the flight path to be a curved line. Direction of wind drift may be determined by noting any change in magnetic bearing while homing with the radio compass. An increasing magnetic bearing indicates a wind from the right, while a decreasing magnetic bearing indicates a wind from the left. Compensate for wind drift by off-setting the airplane heading until there is a minimum rate of change of the magnetic compass reading. The radio compass indicator now shows directly in degrees the relative airplane-to-station heading necessary to correct for wind drift.

POSITION FINDING.

GENERAL.

There are two methods of position finding: automatic and aural-null. Prior to using either method, perform the following steps to shorten the time required for a complete set of readings:

a. Select three stations whose geographical locations are spaced at approximately equal intervals about the airplane.

b. Tune in the stations, identify them and log their dial readings.

AUTOMATIC METHOD.

For operation as an automatic indicating position finder, perform the following operations:

a. Adjust VAR knob on the indicator until its bearing scale at the index is the same as the CORRECTED magnetic heading of the airplane.

b. Set function switch knob to COMP.

c. Tune in one of the selected stations and record the bearing as indicated by the TAIL of the indicator pointer.

d. Repeat step c. for the other stations, in rapid succession, while flying with a steady level heading.

Note

Because of the airplane's motion, the less time taken for observations, the greater accuracy of the fix.

e. The recorded bearings will be the station-to-airplane bearings from the north. Project lines from the stations at the recorded bearings. The airplane position will be within the vicinity of the small triangle made by the intersection of the projected lines.

AURAL-NULL METHOD.

For operation as an aural-null position finder, perform the following operations:

a. Adjust the VAR knob on the indicator until its bearing scale at the index is the same as the CORRECTED magnetic heading of the airplane.

b. Set function switch to LOOP.

c. Tune in desired station. To obtain good signal strength for station identification, it may be necessary to rotate the loop by means of the LOOP L-R switch knob for maximum signal. Direction and speed of loop's rotation are controlled by direction and amount of LOOP L-R switch rotation, respectively.

d. Use the LOOP L-R switch knob, as in step c., and rotate loop for minimum headset volume. Record the bearing shown by the indicator pointer. Better definition of the null may be obtained by turning the AUDIO control fully clockwise and locating the null by either listening for minimum audio signal or noting a counterclockwise dip of the tuning meter pointer. The use of cw operation also improves the definition of the null. To obtain cw operation, set the CW-VOICE switch to CW position.

e. Position finding in LOOP operation is subject to a 180 degree error, since there are two null points in a 360 degree rotation of the loop. This ambiguity is overcome by keeping aware of your general geographical location and by selecting stations located well to the left and right of your course. Figure 4-3 demonstrates a flight on a north course with station A supposedly dead ahead. If the indicator shows a bearing of 180 degrees when station A is used, you apparently have passed station A and are flying away from it. This indication also could be the result of using the wrong null. However, you know station B is to your left and station C is to your right, so by taking null point bearings on stations B and C, projecting lines through the stations at angles indicated by their bearings relative to north until they cross, you determine your approximate position. You then know whether you actually have passed station A or whether you have used the wrong null.

RECEIVER OPERATION.

ANTENNA RECEPTION.

a. Turn function switch to ANT.

b. Turn band switch to desired frequency band.

c. Set CW-VOICE switch to CW position for aural reception of unmodulated signals.

d. Use TUNING crank and tune in desired station.

e. Adjust AUDIO control for desired headset volume.

Note

For best definition of radio range stations, adjust AUDIO control for lowest usable headset volume and continue to reduce volume as the A-N signals increase in strength.

LOOP RECEPTION.

If reception on ANT. is noisy due to static, better results may be obtained by operating in LOOP position as follows:

a. Turn function switch to LOOP.

b. Turn band switch to desired frequency band.

c. If station is unmodulated, set CW-VOICE switch to CW.

d. Tune in desired station.

e. Rotate loop with LOOP L-R switch until maximum signal is obtained. If flight course is not straight, readjustments may be necessary.

f. Adjust AUDIO control for desired headset volume.

g. For best definition of radio range A-N signals on LOOP, it is necessary to maintain the loop near the 90 or 270 degree position and adjust the AUDIO control for lowest usable headset volume.

Note

Cone of silence indications are not always reliable while receiving on LOOP. In some cases, an increase instead of a decrease in signal may be noted. This is the result of certain types of radio range transmitting antennas and the loop location on the airplane.

SUMMARY OF PRECAUTIONS DURING OPERATION.

a. Select radio stations that provide stable bearings. Do not use a station for bearing unless it can be identified by headset signal on COMP. operation. High powered, clear channel stations should be used when possible. Any interference from other stations will cause an error in bearing. Tune equipment accurately. Station identification must be checked, especially stations broadcasting network programs. Avoid taking bearings on synchronized stations except when close to desired station. If station stops transmitting or fades, bearings may change to other stations of the same frequency, thus causing errors. This is especially true of code stations operating in a network.

b. Night effect or reflection of radio waves from the sky may be recognized by fluctuations in bearings. Night effect is worse at sunrise and sunset. The higher the frequency of operation, the greater the night effect. It may be present at distances over 20 miles when receiving 850 to 1750 kilocycle stations; however, with 100 to 450 kilocycle stations, reliable bearings above 200 miles can be taken even when night effect is present. The remedies for night effect are: increase altitude, thereby increasing signal strength of direct waves; use stations operating on lower frequency and take an average of the fluctuations.

c. Mountain effect is considered to be the reflection of radio waves from mountain surfaces. It is known to exist around Salt Lake City and Pittsburgh. Do not rely on bearings taken in such areas.

d. For aural reception of A-N signals, operate equipment on ANT. or LOOP instead of COMP., since the action of AVC in COMP. position will cause broad course indications. Always operate the equipment with AUDIO control set at lowest usable headset volume and reduce it as the A-N signal strength increases. Cone of silence indications are not always reliable when operating the equipment on LOOP. Use equipment on ANT. for cone of silence indication.

e. This equipment should provide compass bearings during conditions of moderate static which interrupt normal reception. When static becomes too severe, it will be necesary to operate on LOOP position. In this position, satisfactory aural reception and aural-null direction finding will be possible most of the time.

f. Do not depend on two stations for fix of location. Use at least three stations, with bearings spaced at approximately equal intervals throughout 360 degrees, for greatest accuracy.

g. While taking bearings, always keep airplane on a steady level heading.

h. When homing or direction finding on LOOP operation, there is a 180 degree ambiguity, and station bearings may be 180 degrees from the null obtained. Use stations with good signal strength for sharply defined nulls. Width of null may be controlled by position of AUDIO control. The tuning meter may be used as a visual-null indicator.

AN/ARR-2A—NAVIGATION (HOMING) RECEIVER (R-4A/ARR-2).

a. On NAVIG console C-738/ARR-2 control unit, set PITCH selector switch to NAV and turn CHAN SEL to desired channel. Advance SENS control to produce a usably weak signal or, if desired signal cannot be heard, to produce a fairly strong background noise. When signal is received, adjust PITCH control to produce a pleasing audible tone. Readjust SENS control to obtain a usably weak signal. The secret of accurately interpreting navigation signals lies in using the lowest satisfactory setting of SENS control. Keep this control adjusted to receive only one character predominantly.

b. At completion of this check, turn SENS to minimum.

c. The volume control on the master control unit C-739/ARC does not affect the volume of the navigation receiver.

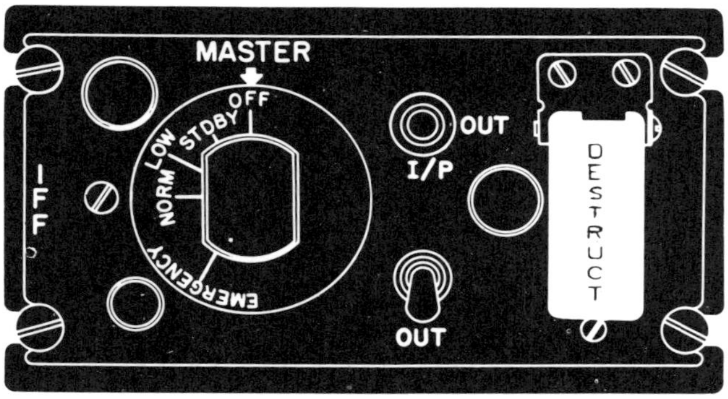

Figure 4-4. IFF Control Panel

Note

When AN/ARC-27 is installed, the C-739/ARC master control is omitted. In this configuration the AN/ARR-2A equipment is connected directly to the power source and may be turned off only by opening the circuit breaker.

SIMULTANEOUS OPERATION OF RECEIVERS.

One or more receivers of AN/ARN-6, AN/ARR-2A, AN/ARC-1 or AN/ARC-27 or -27A, and AN/ARA-25 may be monitored simply by advancing the associated volume or sensitivity control.

AN/APX-6—IFF EQUIPMENT.

PURPOSE.

The radar identification set AN/APX-6 is an airborne transponder and is one of several equipments which may be operated together to provide a system of electronic identification and recognition. The purposes of the AN/APX-6 are:

a. To identify the airplane in which it is installed as friendly when it is correctly challenged by friendly shore, shipboard or airborne radars.

b. To permit surface tracking and control of airplane in which it is installed.

OPERATION.

All controls required for operation of AN/APX-6 equipment are located on radar set control C-629/APX-6. This unit is located on the radio switch panel on the right console.

Operation is as follows:

a. To turn equipment on, rotate master selector to NORM.

b. To indicate emergency or distress, press red dial-stop (see figure 4-4) and rotate master selector to EMERGENCY.

Note

The following is a suggested technique for right hand, rapid operation of the master switch to obtain emergency operation: grasp the knob between the knuckles of the first and second fingers, at the same time pressing the dial stop with the thumb, and rotate the switch to EMERGENCY position.

c. To maintain the equipment ready for instant use, but inoperative, rotate the master selector to STD-BY.

d. The detent position labeled LOW on the master selector should not be used except upon proper authorization.

e. The switches labeled MODE should be set to their OUT positions unless otherwise directed by proper authority.

f. To turn off the equipment, rotate the master selector switch to OFF.

AN/APN-1—RADIO ALTIMETER.[1]

The radio altimeter equipment consists of a receiver-transmitter mounted on the radio deck in the fuselage, an altitude limit switch (9, figure 1-3) located on the left console, an altimeter (9, figure 1-3A or 12, figure 1-3D) located on the left side of the instrument panel with an adjacent altitude low limit indicator light, and two antenna assemblies. The equipment is designed to provide direct measurement of altitude relative to the terrain during flight within a low range of 0 to 400 feet and a high range of 400 to 4000 feet. When operating in the high range, the altitude limit switch settings are ten times those shown on the dial.

AN/APN-22—RADIO ALTIMETER.[2]

The radio altimeter equipment consists of a radar receiver-transmitter, an electronic control amplifier, and a height indicator. This equipment provides reliable altitude measurements of 0 to 10000 feet over land and 0 to 20000 feet over water.

The altitude of the airplane is shown on a single-turn type indicator mounted on the upper left side of the main instrument panel.

The drop-out altitude (the altitude at which the signal becomes too weak to operate the system) is above 10000 feet over land and higher when over water.

Note

The drop-out altitude is reduced in banks, and climbs or dives of approximately 60 degrees.

A reliability circuit disables the indicator when the signal is too weak, and the pointer will assume an off-scale position behind a mask. There is an identical indicator in the photo recorder bay.

The radio altimeter uses both ac and dc. The 28 volt dc for the equipment is supplied by the main bus through the radio altimeter circuit breaker. The 115 volt ac for the system comes from the No. 2 inverter, which obtains the necessary dc from the main bus. The a-c

[1] F9F-6P airplanes ser No. 127473 through 127488.

[2] F9F-6P airplanes ser No. 127489 and subsequent.

circuit from the inverter is protected by a circuit breaker and a fuse.

CONTROL.

The only pilot operated control on this equipment is the on-limit knob located on the lower left corner of the indicator bezel. This control operates the system on-off switch and is also used to select the limit altitude. The adjustable index at the outside of the scale can be preset to the limit altitude by turning the on-limit knob until the index is over the desired altitude on the scale. The altitude of the airplane can then be maintained by observing the position of the pointer with respect to the index, without considering the actual scale calibrations. The indicator has a low limit light mounted in the lower right corner of the bezel. This light will flash red when flying at or below the altitude to which the index has been preset.

OPERATION.

a. Turn equipment on by rotating on-limit control clockwise. Wait one minute for equipment to warm up.

b. Turn on-limit control until index indicates desired altitude limit. The low limit warning light will flash on when the airplane is flying at or below the altitude indicated by the index.

c. Turn equipment off by rotating on-limit control knob counterclockwise as far as it will go.

LIGHTING EQUIPMENT.

EXTERIOR LIGHTING SYSTEM.

The exterior lighting system receives power from the main bus and is controlled by switches grouped on the exterior lights control panel on the right console (3, figure 1-3B). Toggle type selector switches with the positions BRIGHT, OFF, and DIM are provided for the formation, fuselage, tail position, and wing position lights. The master switch controls the exterior lights for manual, code, flash, or steady operation, depending on switch position. A code selector switch permits automatic signaling in code with the fuselage lights. Manual signaling is accomplished by a keying switch. The flasher-coder unit performs the dual function of alternately flashing the wing position fuselage and tail lights and automatically flashing the fuselage lights in code according to the letter selected on the code selector switch. When the exterior lights master switch is in any on position, and the wheels are down and locked, the flasher-coder will automatically flash the approach light if the arresting hook is not down. Circuit protection is afforded by two EXT. LIGHTS circuit breakers located on the main circuit breaker panel and fuses on the right console.

CONTROLS.

EXTERIOR LIGHTS CONTROL PANEL.

The exterior lights control panel is installed on the right console. The panel contains a master switch, a code selector switch, a keying switch, an indicator light, and individual light switches for selective control of the formation, fuselage, and wing position light circuits.

EXTERIOR LIGHTS MASTER SWITCH.

The exterior lights master switch is located on the exterior lights control panel and is of the rotary type, with OFF, MAN, CODE, FLASH, and STDY positions. The OFF position deenergizes all exterior light circuits. When the switch is set to STDY, all exterior lights are illuminated continuously, provided the associated light switches are turned on. Setting the master switch to FLASH permits the wing tip position lights and white tail position light to flash alternately with the fuselage lights and the yellow tail position light. The CODE position is used in conjunction with the code selector switch to flash the fuselage lights in code automatically according to the letter selected. The MAN position permits manual signaling with the fuselage lights by operating the keying switch. Visual indication of the manually keyed signal is provided by a red indicator light adjacent to the keying switch on the control panel.

CODE SELECTOR SWITCH.

The code selector switch is located on the exterior lights control panel and is used to flash out in code on the fuselage lights any one of the 12 letters of the alphabet that is selected. It is used only when the master switch is set at CODE.

CODE KEY.

This key is located on the exterior lights control panel. When the master switch is set at MAN, code messages can be tapped out on the fuselage lights by using this key, provided the fuselage lights selector switch is set to DIM or BRIGHT.

CODE KEY INDICATOR LIGHT.

The red indicator light on the exterior lights control panel flashes the signal being transmitted by the code key on the fuselage lights.

FUSELAGE LIGHTS.

The two fuselage lights are installed on the top and the bottom of the fuselage approximately midway between nose and tail. Each light contains a normal and a dimming lamp. The lights are controlled by the fuselage lights selector switch and the master switch.

FUSELAGE LIGHTS SELECTOR SWITCH.

The fuselage lights selector switch is a three position toggle switch located on the exterior lights control panel. It can be set to BRIGHT, DIM or OFF. When the master switch is set to STDY and the fuselage light selector switch is set to DIM or BRIGHT, the fuselage lights will burn continuously. When the master switch is set to FLASH and the fuselage lights selector switch is set to DIM or BRIGHT, the fuselage lights and the yellow tail position light will flash alternately with the wing position lights and white tail position light. When the master switch is set to CODE, the fuselage lights will automatically signal in code the letter selected on the code selector switch. If the master switch is set to MAN, the fuselage lights will operate in conjunction with the code key for manual signaling purposes.

WING POSITION LIGHTS.

The recessed type wing position lights are located on the leading edge of each wing tip. The right wing position light cover is green and the left is red. These lights are controlled by the wing position light selector switch and the master switch located on the exterior lights control panel.

WING POSITION LIGHTS SELECTOR SWITCH.

This three position toggle switch, which can be set to BRIGHT, DIM or OFF, is located on the exterior lights control panel. When the master switch is set to MAN, CODE, or STDY and the selector switch is set to DIM or BRIGHT, the wing position lights will burn continuously. The wing position lights and the white tail position light will flash alternately with the fuselage lights and yellow tail position light when the master switch is set to FLASH.

TAIL POSITION LIGHTS.

The tail position lights are installed on the end of the tail fairing. The upper light is fitted with a yellow cover, the lower with a white cover. These lights are controlled by the tail position selector switch and the master switch.

TAIL POSITION LIGHTS SELECTOR SWITCH.

This toggle switch can be set at any of three positions, BRIGHT, DIM or OFF, and is on the exterior lights control panel. When the tail position lights selector switch is set to BRIGHT or DIM, the tail position lights will burn continuously if the master switch is set to MAN, CODE or STDY. If the master switch is set to FLASH, the yellow tail position and fuselage lights will flash alternately with the white tail position and wing position lights.

FORMATION LIGHTS.

A white formation light is located on the underside of each outboard wing panel near the tip. The lights are controlled by the four positions of the master switch and by the formation light selector switch.

FORMATION LIGHT SELECTOR SWITCH.

The formation light control is a three position toggle switch on the exterior lights control panel with BRIGHT, DIM and OFF positions. The formation lights will glow continuously when the master switch is set to MAN, CODE, STDY or FLASH and the formation light switch is set to BRIGHT or DIM .

INTERIOR LIGHTING SYSTEM.

The interior lighting system consists of an edge lighted instrument panel, certain edge lighted console control panels, and two hooded lights on the left side of the cabin and two on the right side to illuminate the consoles. Power is supplied from the main bus and circuits are protected by two circuit breakers on the main circuit breaker panel (1, figure 1-3B) and two fuses mounted on the fuse panel (8, figure 1-3B). The interior lighting system is controlled by two rotary switches (5, figure 1-3B) on the right console.

INTERIOR LIGHTS CONTROL SWITCHES.

The instrument panel edge lighting is controlled by a combination on-off switch and rheostat mounted on the right console. The edge lighted console control panels and console lights are controlled by a switch identical to the instrument panel edge lighting switch and mounted adjacent to it. Turning the switches clockwise from off turns on the respective lights and provides a range of adjustment from dim to bright.

SPARE LAMPS.

Spare lamps are mounted on a panel on the right console (10, figure 1-3B).

OXYGEN SYSTEM.

DESCRIPTION.

Two 514 cu in. capacity, shatter-proof oxygen cylinders are installed in brackets on the bulkhead aft of the cabin. The cylinder valves are opened when connected to the line coupling, and flow is controlled by the regulator. The diluter-demand pressure breathing regulator, Pioneer Model 2867-A1, with oxygen supply shut-off valve, diluter valve control, flow indicator and system pressure gage, is installed on the left console forward of the throttle quadrant. The feed tube from the regulator runs to a personal gear composite disconnect on the aft end of the console, and a type A-13a face mask attaches to the tube from the assembly (1, 17, figure 1-3).

OXYGEN REGULATOR OPERATION.

The automatic pressure breathing diluter-demand regulator is designed to meet the demands of the inhalation phase of the breathing cycle and deliver either a properly proportioned mixture of air and oxygen or 100% oxygen, dependent upon the setting of the adjustable diluter valve lever. With the diluter valve set to the NORMAL OXYGEN position, air is drawn into the breathing system and is automatically mixed with oxygen from the supply cylinder to give the total needed oxygen required up to approximately 30000 feet, beyond which 100% oxygen is delivered. With the diluter valve set to the 100% OXYGEN position, 100% oxygen is delivered at all altitudes. Above 35000 feet cabin pressure, the regulator automatically delivers to the mask the required positive pressures of oxygen for operational flights up to an equivalent cabin pressure of 43000 feet and may be used for short periods of time at an ambient pressure of 45000 feet. In addition, sufficient pressure is delivered automatically for emergency descent from 50000 feet in the event that pressurization is lost and the cabin pressure is reduced to that of the surrounding air. With the diluter valve set to the NORMAL OXYGEN position, a relatively small inhalation suction (one inch of water suction) is suffi-

Oxygen Duration - Hours

CABIN PRESSURE ALTITUDE	5000 feet	10000 feet	15000 feet	20000 feet	25000 feet	30000 feet
Hours Duration Using Diluted Oxygen	15	18.5	17	13.8	8	6.2
Hours Duration Using 100% Oxygen	1.9	2.4	2.9	3.8	4.6	6

Note

Duration data should be used as a guide only, since oxygen consumption varies with the individual.

Figure 4-5. Oxygen Duration Chart

cient to deliver a flow of 150 liters of oxygen per minute. This characteristic assures the user an adequate oxygen flow and ease of breathing. The regulator is attached directly to the high pressure oxygen supply through 3/16 inch OD copper tubing connected to the cylinders. The pressure in the cylinders may decrease from 1800 psi to 50 psi without affecting the normal operation of the regulator.

OXYGEN PREFLIGHT CHECK.

The following items shall be checked prior to a flight in which oxygen is to be used or is likely to be used, in order to assure proper functioning of the system:

a. After the oxygen supply control at the regulator has been turned on, oxygen supply cylinder pressure should read 1800 to 1850 psi if the cylinders are fully charged. If the cylinder pressure has decreased by more than 25 pounds in 24 hours, the leakage is excessive and the system should be subjected to ground crew test prior to use.

Oxygen regulator leakage. After determining that cylinder is fully charged, close the shut-off valve. If the pressure gage shows a drop of not more than 300 psi in four minutes, leakage is acceptable; turn on oxygen supply valve and continue the tests.

b. To test the breathing tube couplings, regulator diaphragm, and diluter check valve for leakage, insert a spare mask tube quick disconnect fitting into the open end of the disconnect. Blow into the open end of the disconnect until the flow indicator face opens. Seal the end of the disconnect with the tongue. If the flow indicator does not close within five seconds, the leakage is within acceptable limits. If leakage exists, check the couplings, outlet elbow and breathing tube hose clamps for tightness.

c. Put on type A-13a mask. Check mask fit by placing the thumb over the disconnect at the end of the breathing tube and inhaling lightly. The thumb must be removed from the disconnect after each continuous inhalation. If there is no leakage, the mask adheres to the face tightly and a definite resistance to inhalation will be encountered. If the mask leaks, tighten the mask suspension straps.

WARNING

Do not use a mask that leaks.

d. Fully engage mating portion of the disconnect coupling to connect mask to the oxygen system breathing tube.

e. Attach clip of breathing tube to proximate strap of shoulder harness sufficiently high on the chest to permit free movement of the head without stretching the mask tube.

f. Breathe several times and observe flow indicator for blink, verifying the positive flow of oxygen.

OXYGEN SYSTEM OPERATING INSTRUCTIONS.

a. Oxygen shall be used constantly during day flights when above 10000 feet and during night flights when above 5000 feet and when on combat missions and training missions simulating combat.

b. After making certain that oxygen supply control lever is turned to open supply to regulator, see that pressure gage reads approximately 1800 psi for fully charged cylinders.

c. Set diluter valve to NORMAL OXYGEN for normal flight conditions. Set diluter valve to 100% OXYGEN prior to firing guns or when the presence of carbon monoxide or other noxious gases or fumes is suspected.

d. Put on the oxygen mask, fully engaging disconnect coupling. Attach clip of breathing tube to proximate strap of shoulder harness sufficiently high on the chest to permit free movement of the head without stretching the mask tube.

e. To check mask fit during flight, turn the safety pressure control to PRESSURE ON. Take a deep breath and hold breath. Note position of the oxygen flow indicator. If flow indicator opens (i.e., all black), excessive leakage is indicated. Tighten mask straps until flow indicator closes (i.e., white face shows), indicating that a leak-tight mask seal has been obtained. The characteristics of the flow indicator are such that this test cannot be conducted at pressure breathing altitudes; however, the outward flow of oxygen into the eyes from a leaking mask is readily detectable and is an equally sensitive test.

f. Breathe normally.

g. When breathing oxygen during flight, check the following items frequently:

System pressure gage for oxygen supply.

Flow indicator for flow of oxygen through regulator.

Mask fit for absence of leaks.

Assure engagement of mask tube to composite disconnect.

Connection of emergency oxygen fitting to mask-to-composite disconnect coupling.

h. The oxygen flow indicator operates upon the intermittent application of from five to seven inches of water pressure created by the flow of oxygen. The automatic pressure breathing oxygen regulator delivers a pressure of five to seven inches of water to the mask at 37000 to 39000 feet and this pressure is likewise transmitted to the oxygen flow indicator, which will remain open as long as this pressure is applied. Accordingly, the flow indicator will not blink above this loading; however, the positive pressure in the mask is an unmistakable indication that oxygen is being delivered to the mask and no apprehension should be felt as the flow indicator remains open.

Note

Do not exhaust supply cylinders below 300 psi except in an emergency.

EMERGENCY OXYGEN SYSTEM OPERATION.

Should symptoms occur suggestive of the onset of anoxia, immediately turn the safety pressure control on the rim of the regulator to PRESSURE ON and descend below 10000 feet.

WARNING

If regulator becomes inoperative in flight, activate emergency oxygen supply equipment, and descend below 10000 feet.

Whenever excessive carbon monoxide or other noxious or irritating gas is present or suspected, the diluter valve should be set at 100% OXYGEN, regardless of the altitude, and undiluted oxygen used until danger is past or the flight completed. Should brief removal of mask from the face be necessary at high altitudes, use the following procedure:

a. Take three or four deep breaths of undiluted oxygen (diluter valve set at 100% OXYGEN).

b. Hold breath and remove mask from face.

c. As soon as practicable, replace mask to face and take three or four deep breaths of undiluted oxygen.

d. Reset diluter valve to NORMAL OXYGEN.

Above 35000 feet cabin altitude, the regulator automatically delivers to the mask the required positive pressures of oxygen for operational flights up to an equivalent cabin pressure of 43000 feet and may be used for short periods of time at an equivalent cabin pressure of 45000 feet. In addition, sufficient pressure is delivered automatically for emergency descent from 50000 feet in the event that pressurization is lost and the cabin pressure is reduced to that of the surrounding air. Upon completion of a flight, set diluter valve to 100% OXYGEN.

NAVIGATION EQUIPMENT.

ELECTRONIC NAVIGATION EQUIPMENT.

RADIO COMPASS AND NAVIGATION (HOMING) RECEIVER.

See applicable paragraphs under Electronic Equipment.

NAVIGATION INSTRUMENTS.

STAND-BY COMPASS.

See Instruments paragraph, Section I.

G-2 COMPASS.

See Instruments paragraph, Section I, for description.

OPERATION.

a. Set battery switch to BAT. & GEN.

b. Allow sufficient time for the gyro motor to warm up, then set control switch to COMPASS CONTROL.

c. Push in and rotate caging knob to establish initial "on course" heading obtained from correspondence indicator, then pull out knob. The instrument will now correct any deviation from the magnetic heading, as shown by the correspondence indicator, at a rate of four degrees per minute.

CAUTION

To ensure proper alignment of the gyro element, it is important that the knob be held firmly against the cover flange while it is being rotated and also for a few seconds after the correct setting has been made. Release the knob straight out—with no twisting movement.

d. When operating near polar regions or on a carrier deck, the instrument should be used as a free directional gyro; to so use, set control switch to FREE DG. The gyro drift should be checked every 15 minutes and reset if necessary.

e. To return to compass controlled operation from a free directional gyro, set control switch to COMPASS CONTROL, then push in caging knob and reset master direction indicator to heading indicated by correspondence indicator.

CAUTION

Always operate the selector switch before setting the heading with the caging knob, otherwise the instrument will be damaged.

If the difference between the correspondence indicator reading and the gyro reading is great, several minutes will be required before complete correction of the instrument will take place. To eliminate this delay, the gyro and correspondence indicator may be synchronized by turning the caging knob. Caging is not required during maneuvers.

The G-2 compass gyro is of a nontumbling type which does not require caging. This, however, does not preclude a caging operation upon completion of flight maneuvers in order to override the normally slow gyro erection and correction rates incorporated in the system.

ARMAMENT.

GENERAL.

Four M-3 20mm guns are installed in the fuselage nose. Ammunition is loaded in four ammunition boxes connected to the guns by feed chutes. An Aero 5A armament control system is installed and an AN-N-6A gun camera is mounted on the windshield deck. Armor plate is installed on the bulkhead forward of the pilot's seat and on the sloping bulkhead aft of the seat. The bullet resistant glass windshield is mounted in a heavy aluminum alloy plate frame.

Note

On flights in which no ammunition, or ammunition for only two guns is loaded, ballast must be installed in order to maintain airplane cg within the aft limit at low fuel loads. This ballast is provided for the airplane by the contractor and is stowed in the forward (inboard guns) ammunition boxes. On flights in which full ammunition for all guns is loaded, the cases and links from fired rounds are retained in the nose shell, thereby maintaining cg within proper limits.

GUN CONTROLS.

The trigger switch (14, figure 1-4) is on the front of the stick grip. The four gun chargers are controlled by a four way, solenoid operated valve and pressure switches which, in turn, are controlled by two toggle switches on the armament switch panel, located on the windshield deck to the left of the sight unit. These switches control the inboard and outboard guns and have positions marked READY-OFF-SAFE. The armament master toggle switch is outboard of the other switches on the armament switch panel (10, figure 1-3A).

GUN CONTROLS OPERATION.

READY SETTING.

When either the inboard or outboard guns are selected for firing, set the master switch to ON, set the applicable gun charging switch to SAFE, and then set it back to READY. This operation will permit hydraulic pressure to act on the charger pistons, which retract the breechblocks to the seared position. After the breechblocks have been seared, the fluid on the forward side of the piston in the cylinders returns to the reservoir and the gun charger piston returns to the battery position, rendering the guns ready for firing.

SAFE SETTING.

When either READY-OFF-SAFE switch is set to SAFE, the respective breechblocks are moved automatically from battery to recoil and are retained in the recoil position by the sear and charger action so that the trigger controls will not operate even if the trigger is depressed. The breechblocks will remain in recoil, preventing firing, until either switch is set to READY, which releases the hydraulic pressure and permits the charger piston to return to the battery position and permits firing if the trigger button is pressed.

GUN OPERATING PROCEDURE.

a. Set armament master switch to ON.

b. Set selected READY-OFF-SAFE (outboard, inboard, or both) switch to SAFE and then back to READY for firing.

c. Set diluter valve lever to 100% OXYGEN prior to firing.

d. Press trigger switch on stick grip. See Section IV of CO 01-85FGD-1A for additional gun firing information.

e. After firing, return READY-OFF-SAFE switch, or switches, to SAFE.

Note

To operate the guns for ground check, the disabling switch on the right side of the nose wheel well must be set to energize the gun circuit, which is normally broken when the landing gear control handle is at DOWN.

In order to ensure that the gun charger lug has fully returned to the battery position prior to actuating the trigger button for firing of the guns, the electrical circuit between the trigger button and the gun-firing solenoid remains open until the hydraulic pressure in the gun charging cylinder drops to the low pressure setting on the pressure switch. At high altitudes where the temperature is −40°C (−40°F) or lower, it takes about three seconds to close the circuit to the gun firing solenoid. Whenever the pilot presses the trigger but-

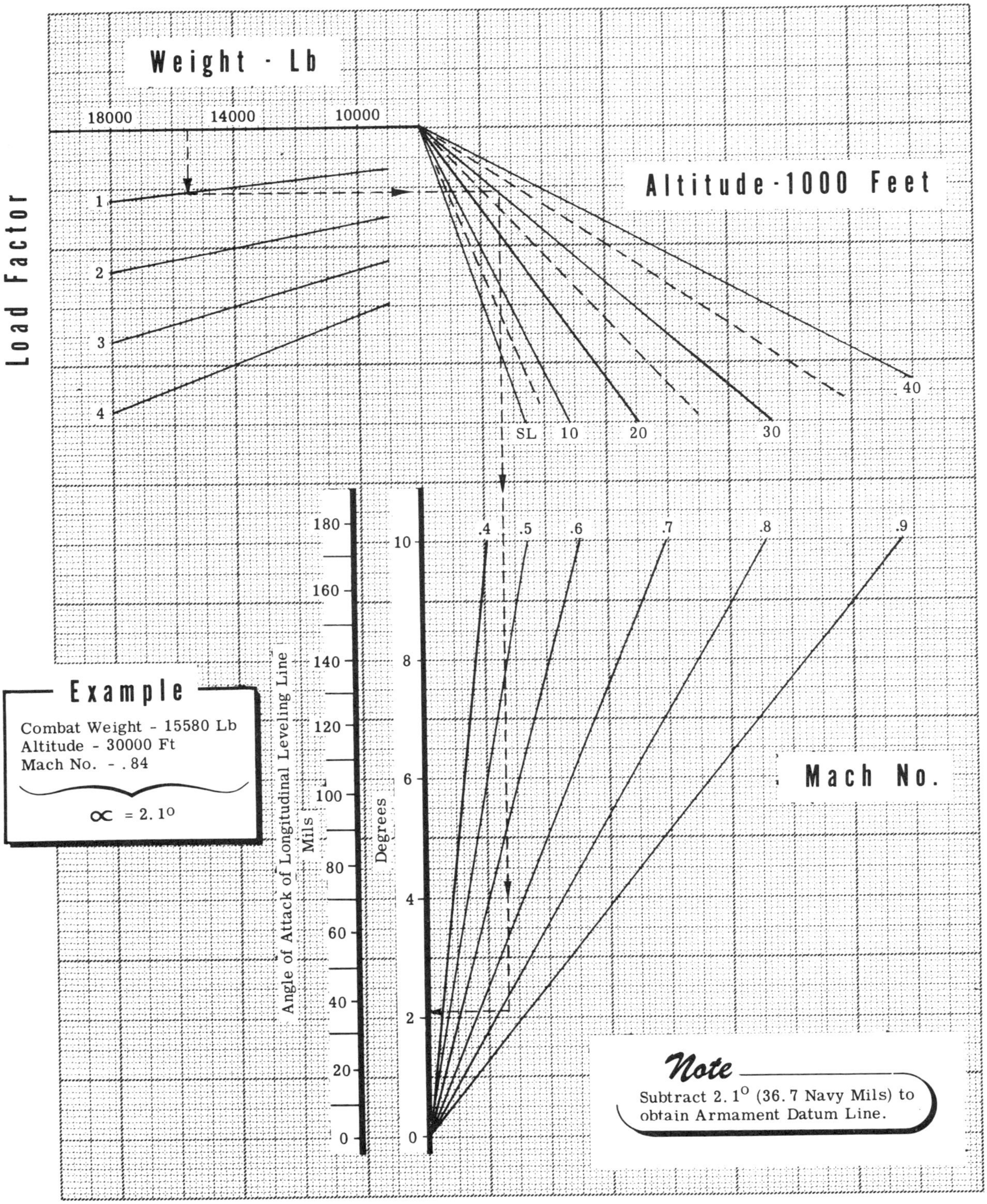

Figure 4-6. Angle of Attack Variation with Speed Weight and Normal Acceleration

ton after actuation of the READY-OFF-SAFE switch to READY and the guns fail to fire, it may be an indication that the charger lug has not returned to battery position. If such is the case, keeping the button depressed for several seconds will allow the charger lug to return to battery position and start firing of the guns.

WARNING

The guns will be operative when switch is set to READY.

ARMAMENT CONTROL SYSTEM—AERO 5A.

The ACS Aero 5A comprises a sight unit MK 8 Mod 0 (a part of AFCS MK 6 Mod 0), mounted on a bracket on the cowl center line (11, figure 1-3A; an Aero 2C gunsight controller, located below the sight; a MK 20 Mod 0 control box on the left console (11, figure 1-3); an Aero 4C ranging control (the throttle grip) (22, figure 1-3); a MK 7 Mod 0 relay box; a radar range servo Aero 1A and an Automatic-Range-Only (ARO) radar set AN/APG-30. The radar set control C-775/APG-30, is located on the left console (12, figure 1-3) adjacent to the MK 20 Mod 0 control, and the ID-295/APG-30 tracking indicator is mounted on a bracket (12, figure 1-3A) attached to the sight unit MK 8 Mod 0. This unit contains a red tracking indicator light, a range meter and an automatic frequency control meter. The red tracking indicator light will flash on while the radar is searching for a target or is radiating signals that might be detected by enemy countermeasures, and will go out when the radar has locked on a target. The range meter indicates the range of the target in feet, and the automatic frequency control meter measures crystal current in milliamperes. The latter instrument is used to check that the system is operating properly.

CAUTION

The RANGE and AFC meter dials have been rotated 90 degrees from the normal positions to eliminate magnetic interference with the stand-by compass.

On certain airplanes,[1] an armament selector panel is installed on the left console. This panel contains a two position guns and rockets selector switch marked GUNS and ROCKETS, a two position dive angle selector switch marked 35° & OVER and 35° & UNDER, a three position rocket arming switch marked ARMING NOSE & TAIL, SAFE, and TAIL ONLY, and a three position, rotary type rocket station selector switch marked LEFT, RIGHT, and BOTH.

SYSTEM CHECK.

a. Set battery switch to BAT. & GEN.

b. Set power switch on C-775/APG-30 control to STDBY MAN. Check that amber light is on.

Note

Placing the radar in STDBY MAN. permits it to warm up and also permits manual ranging of the fire control system.

c. Rotate selector switch on MK 20 Mod 0 control to GYRO. The gyro motor should then start to run, and the gyro reticle image should appear on the reflector plate of the sight unit. The image should stabilize quickly and be clearly defined.

d. Rotate the selector switch to FIXED & GYRO and note that the fixed image of the sight unit is visible.

e. Rotate dimmer control knob on the MK 20 Mod 0 control from DIM to BRIGHT, noting that the dimmer control varies the intensity of the images continuously over the entire range.

f. Rotate the selector switch to the FIXED position. The gyro image should then disappear and the gyro motor come to a gradual stop.

g. Return the selector switch to the FIXED & GYRO position and rotate the throttle grip (ranging control) to its counterclockwise stop, then hold. The range scale in the sight unit should indicate 600 feet.

h. Note diameter of circle formed by the diamond-shaped pips of the gyro image, then rotate throttle grip full clockwise. The range scale in the sight unit should now indicate 2400 feet and the diameter of the circle should be smaller.

Note

Step i. below applies only if AN/APG-30 radar equipment is not being used.

i. Move the target span handle on the sight unit from right to left toward the higher setting on the span scale. The diamond-shaped pips should then move out from the center pip of the gyro image. The sight line through the center pip of the gyro reticle should be the same as that through the fixed reticle.

j. Set the guns and rockets selector switch to GUNS.

k. Sight on a distant object. Both the fixed and gyro images should appear to coincide on the object. Rotate ranging throttle grip from maximum to minimum range. The center pip of the gyro image should remain stationary. A slight vertical movement of the center pip may be found in some sight units when range is changed from maximum to minimum. This is acceptable if the total movement does not exceed four mils (this may be estimated by comparison with the ten mil arcs of the fixed images), and if the center pip coincides with the center of the fixed cross at some range between maximum and minimum.

l. Set the guns and rockets selector switch to ROCKETS.

[1] F9F-6 airplanes ser No. 128155 and subsequent.

m. Set the dive angle switch to 35° & UNDER. The center pip of the gyro image should move to a point which corresponds to the offset value specified for rocket dive angles from 15 to 35 degrees. The ten mil arcs of the fixed image may be used to gage the amount of offset from the center of the fixed cross.

n. Set the dive angle switch to 35° & OVER. The center pip should move to a point which corresponds to the offset value for rocket dive angles from 35 to 60 degrees. Check the caging circuit by changing the position of the dive angle switch while watching the pip. With the caging switch closed (throttle grip at extreme maximum range position), the pip should move to its new position. With the caging switch open, the movement should be slow.

o. If the flight is to begin immediately, leave the selector at the FIXED & GYRO setting; otherwise, return the switch to the OFF position.

p. When the AFC meter on the ID-295/APG-30 stops sweeping and drops to zero, place the power switch on the C-775/APG-30 control to ON. The AFC meter will jump to a reading between 0.5 and 1.0 ma and will hold steady. Check to see that the green light on the control is on. The ID-295/APG-30 tracking indicator will flash on while the radar is searching a target in range, and will go out when the radar has locked on a target.

Note

When the power switch on the C-775/APG-30 control is set to STDBY MAN during warm-up (approximately four minutes), the RANGE meter on the ID-295/APG-30 tracking indicator will sweep and the AFC meter will jitter. After warm-up is completed, there should be no movement on the RANGE or AFC meters. The AFC meter should indicate crystal current from 0.5 to 1.0 milliampere, with 0.7 milliampere desired. When the power switch is set to ON, the RANGE meter will sweep until locked on a target. When locked on, the meter will indicate target range in feet.

The radar will lock on the nearest target within its range and show distance to the target, up to 2400 feet, on the sight unit range scale and up to 12000 feet on the ID-295/APG-30. Depressing the range gates switch to GATES CUT will cause the radar to unlock from this target and select another target farther away. If no other target is within the maximum range of the radar, it will sweep to its limit and again select the closest target within its range. Also note movement of the diamond-shaped pips as targets at different ranges are selected.

Note

The radar range exceeds the sight unit range; therefore, the range scale on the sight unit will indicate 2400 feet for all targets beyond this range. The range meter on the ID-295/APG-30 will indicate target range continuously.

FLIGHT ADJUSTMENTS.

To use the system accurately and most effectively at the beginning of the flight:

a. Set power switch on C-775/APG-30 control to STDBY MAN.

b. On MK 20 Mod 0 control, turn selector switch to FIXED & GYRO if desired to see both image patterns; however, if they are confusing, setting selector to GYRO will eliminate the fixed image, or the outer markings of the fixed image can be blanked out by pushing the masking lever forward.

c. Adjust dimmer control knob on MK 20 Mod 0 control until reticle image is at desired brilliance.

d. As soon as the target is sighted, turn power switch on C-775/APG-30 control to ON. If AN/APG-30 radar equipment is not being used, set the target span into the sight unit with the span setting handle (C-775/APG-30 in STDBY MAN). This setting will be a selected dimension of the target airplane, usually its wing span in feet.

WARNING

When no target is in view, keep the power switch in STDBY MAN. Otherwise, radiation of the radar transmitter may give your position away to enemy countermeasures.

Note

Adjustment of target span handle is required only when manual ranging is used.

e. Since the ACS Aero 5A does not compensate for gravity drop, it is necessary to make an allowance of about 16 feet (7 mils) at 2400 feet range, decreasing to about one foot (2 mils) at 600 feet.

Note

The maximum range control on the C-775/APG-30 control is used to adjust the maximum range of the radar. The proper setting of this control will depend on the altitude of the airplane and should be adjusted so that the radar will not lock on ground reflections.

With control set at minimum, the radar range will still be more than the sight unit range. In normal operation at altitudes over 10000 feet, the maximum range control may be left at the full clockwise position to allow locking on a target at maximum range.

When there is more than one target close to the radar line of sight, the desired target may be selected by operating the range gates switch to GATES OUT and noting the change in range shown on the ID-295/APG-30. When the targets are within range of the sight unit (2400 feet), note the change in diameter of the diamond-shaped pips.

Commence tracking by maneuvering the airplane to bring the center pip exactly above the target by an amount necessary to compensate for gravity drop.

When Aircraft Armament Change No. 77 is accomplished on the Radar Range Aero 1A, the pilot may uncage the gyro in the sight unit any time he desires to do so, independent of radar lock-on, by rotating the throttle grip at least 20 degrees out of its detent position. This rotation of the throttle grip when using radar ranging in no way affects the range being fed to the sight unit from the radar. All it does is to uncage the sight unit at the pilot's discretion. The pilot should uncage as soon as possible and establish smooth tracking on the target. The radar will probably lock on in excess of 4000 feet. The sight unit will have a 2400 foot sensitivity from the time of uncaging to 2400 feet. When the range decreases to 2400 feet, the sight will already have solved for the proper lead angle if tracking is commenced at least one second before reaching 2400 feet. The red light in the ID-295/APG-30 will go out when the radar is locked on the target. If the pilot should forget to uncage the sight unit before reaching 2400 feet, the sight unit will automatically uncage at 2400 feet. However, in this case one second of tracking will be necessary before the sight unit generates the proper lead angle. It is therefore always advantageous to uncage the sight unit at least one second before reaching 2400 foot range. When the diamond-shaped pips in the sight unit begin to expand, the range is 2400 feet. Firing may be commenced at this point. When the range decreases to 600 feet the diamond-shaped pips will cease to expand. Since the sight unit will not compute the proper lead below 600 feet, pull-out should commence at or before this point. In addition, the AN/APG-30 radar will probably lose its lock-on at a range of about 675 feet. When this occurs, the sight unit diamond-shaped pips will contract to the 800 yard point. The sight unit will no longer be computing the proper lead.

f. Manual ranging may be selected by turning the power switch on the C-775/APG-30 control to STDBY MAN. After setting the estimated wing span of the target, in feet, into the sight head by means of moving the target span lever to the appropriate number, track during the run by continuously rotating the throttle grip, so as to keep the extreme tips of the target framed by the imaginary circle formed by the inner points of the pips of the gyro reticle image.

Note

On the average airplane, the distance from the junction of the wing root to the end of the tail is half a wing span. At 90 degrees off the tail of the target, ranging is accomplished by keeping the inside tip of one of the pips on the end of the tail of the target and the center pip of the reticle pattern just above the wing root fuselage junction.

During manual ranging, the gyro will remain caged as long as the throttle grip is at the maximum range position (fully clockwise). If tracking is to be started before the target is within range, uncage the gyro by rotating the control slightly counterclockwise.

g. Continue smooth tracking and ranging throughout the attack.

Note

In order to prevent the sight unit gyro from tumbling, it is recommended that the gyro be caged at all times except when tracking and ranging a target.

ROCKET AND BOMB PROVISIONS.[1]

ROCKET CONTROLS.

The rocket firing push button switch is on the side of the control stick grip (17, figure 1-6). The guns and rockets selector two position toggle switch, the dive angle selector two position toggle switch, and the rockets (or bombs) station selector four position rotary switch are mounted on a panel on the left console. The guns and rockets selector switch is marked GUNS and ROCKETS. The dive angle selector switch is marked 35° & OVER and 35° & UNDER. The station selector switch is marked SAFE, LEFT, RIGHT, and BOTH.

ROCKET FIRING OPERATION USING ACS AERO 5A.

To use the ACS Aero 5A as a lead-computing sight when engaging in air-to-ground rocket operations, proceed as follows:

a. On MK 20 Mod 0 control panel, set selector switch to GYRO or FIXED & GYRO and adjust dimmer control to provide proper reticle image contrast for the particular light condition.

b. Set guns and rockets selector switch to ROCKETS.

c. Set dive angle selector switch to 35° & OVER if a steep dive approach to the target is desired, or to 35° & UNDER if a dive angle of 35 degrees or less is best suited for the operation.

d. Set armament master switch to ON.

e. Set station selector switch to LEFT, RIGHT, or BOTH, as operation requires.

Note

Setting this switch to SAFE prevents firing the rockets inadvertently.

f. Cage gyro in sight unit by rotating ranging throttle grip to extreme maximum range position.

g. When target is approached, control airplane so that center pip of gyro image is directly on the target.

Note

The system will compensate for windage and target motion as long as the pilot maneuvers

[1] F9F-6 airplanes ser No. 128155 and subsequent.

the airplane so as to keep the pip on the target. For accurately aiming rockets, it is essential that the pip be held on the target smoothly from the time the approach is initiated until the rockets are released. It is not necessary to range the target by framing it when using the equipment as a lead-computing sight for rocket firing operations.

h. Uncage gyro by rotating ranging throttle grip slightly from extreme maximum range position.

i. After uncaging gyro, hold pip on target smoothly for at least four seconds before firing rockets.

j. Press rocket firing switch on control stick grip to fire rockets.

BOMB CONTROLS.

The station selector switch described in the Rocket Controls paragraph is used in bombing operations, together with a bomb release push button switch on the control stick grip (15, figure 1-6) and an arming selector switch on the left console. The arming selector switch has two positions marked ARMING NOSE & TAIL and TAIL ONLY.

BOMB RELEASE OPERATION.

a. Set armament master switch to ON.

b. Set arming selector switch to ARMING NOSE & TAIL or TAIL ONLY, as operation requires.

c. Set station selector switch to LEFT, RIGHT, or BOTH, as operation requires.

d. Press bomb release switch on control stick grip to release bombs.

BOMBS AND ROCKETS ELECTRIC JETTISON CONTROL.

The bombs or rockets may be jettisoned in an emergency by means of the stores jettison switch on the armament control panel, located on the windshield deck. To release any stores, raise guard and move toggle switch up.

BOMBS AND ROCKETS HYDRAULIC JETTISON CONTROL.

In the event stores cannot be jettisoned by the electric jettison control switch, an emergency hydraulic jettison system is provided. This consists of a "T" handle control, located on the left console, which is pulled aft to actuate a master hydraulic cylinder. Hydraulic pressure from the master cylinder is directed to two slave units (one on each bomb rack) to actuate the stores release mechanisms on the bomb racks.

Note

This hydraulic system is independent of the airplane's hydraulic system.

GUN CAMERA.

The AN-N-6A camera is mounted on a bracket on the windshield deck, forward of the sight unit. Operation is controlled by the armament master switch and the trigger switch on the stick grip. When the guns are fired, the camera operates. The camera may be operated without firing the guns by placing the READY-OFF-SAFE switch in either the OFF or SAFE position and depressing the trigger switch, or by use of a test switch located on the windshield deck.

"G" SUIT EQUIPMENT.

The "G" suit pressure control valve (3, figure 1-3; 2, figure 1-3C) is located on the aft section of the left console. The valve receives air under pressure from the engine bleed air line and meters it to the pilot's "G" suit when a force of approximately 1.75g is applied to the airplane. The control knob, marked HI and LO, allows for adjustment of the rate of inflation of the "G" suit. In the LO range, the valve opens at 1.75g and allows 1.0 psi of air pressure to pass to the suit for every increase of 1.0g force thereafter. In the HI range, the valve also opens at 1.75g but delivers 1.5 psi per 1.0g force thereafter. A button is provided on top of the valve for manually inflating the "G" suit on the ground with the engine running or when in straight and level flight. PRIOR TO EACH FLIGHT, with the engine running and the "G" suit connected, depress this button manually several times to check the system air supply and the operation of the valve. If the valve has any tendency to stick or fails to return to the closed position, it should be replaced. On long flights, this feature makes it possible for the pilot to inflate the suit occasionally for body massage to lessen fatigue. The suit is connected to the pressurized air system at the pilot's gear quick-disconnect receptacle (1, figure 1-3; 1, figure 1-3C), located aft on the left console.

PHOTOGRAPHIC EQUIPMENT—F9F-6P AIRPLANES.

DESCRIPTION.

The nose section of the F9F-6P airplane is divided into a forward and aft camera bay and a viewfinder bay for camera and camera controls installations. The photographic equipment includes cameras, camera mounts, camera controls, a viewfinder and the necessary tubing and wiring. A camera vacuum system and a viewfinder dehydrator system are also included. Camera positioning, camera operation and viewfinder operation are completely automatic and are controlled from the cabin. The camera mounts, with their respective cameras, are installed in the proper bay by using the camera hoisting davit. Access to the camera compartment is through the camera compartment hatch, two camera side windows, and the door and step assembly and adjacent access hole.

TYPE	K-17-6 Inch	K-17-12 Inch	K-17-24 Inch	CA-8	K-18-24 Inch	S-7-S
Focal Length	6 inch	12 inch	24 inch	6 inch	24 inch	6 inch 100 mm
Lens Cone Angle	72°	41°	30°	72°	42°	58° 72°
Shutter Speeds (sec)	1/50 1/100 1/200 1/300	1/75 1/150 1/225	1/50 1/100 1/150	1/100 1/200 1/300	1/50 1/100 1/150	Adjustable Slit
Diaphragm Stops (f value)	f6.3 f8 f11 f16 f22	f5 f6.3 f8 f11 f16 f22	f6 f8 f11 f16	f6.3 f8 f11 f16 f22	f6 f8 f11 f16	Automatic Adjustment
Type of Magazine	A-5-A or A-9	A-5-A or A-9	A-5-A or A-9	MA-6	A-7 or A-8	Component Part of Camera
Negative Size	9 x 9	9 x 9	9 x 9	9 x 9	9 x 18	Strip 9-1/2 inch x 200 feet
Film Capacity	A-5-A - 200 feet 250 exposures A-9 - 390 feet 485 exposures	same	same	250 exposures	A-7 45 exposures A-8 - up to 250 exposures	Continuous Strip
Camera Weight (Loaded)	A-5-A Magazine 71.8 lb A-9 Magazine 78.9 lb	A-5-A Magazine 66.8 lb A-9 Magazine 73.9 lb	A-5-A Magazine 88.8 lb A-9 Magazine 95.9 lb	40 lb	A-7 Magazine 89.7 lb A-9 Magazine 110.4 lb	With Scanner 108.5 lb
Film Loading Note	Total Darkness	Total Darkness	Total Darkness	Total Darkness	Total Darkness	

Figure 4-7. Operational Information—All Cameras

The following items of equipment for photographic reconnaissance may be installed:

Location	*Equipment*	*Type Mount*	*Optical Axis*
Nose Bay	Scanner	Fixed	Vertical
Forward Camera Bay	S-7-S Sonne Continuous Strip Camera	Fixed	Vertical
	or		
	K-17-12 in.	Fixed	Vertical
	or		
	K-17-6 in. or CA-8	Fixed	Vertical
Aft Camera Bay	K-18-24 in. with A-7 or A-8 Magazine	Rotary	3°, 15° and 90° below horizontal
	or		
	K-17-24 in. with A-5 or A-9 Magazine	Rotary	3°, 15° and 90° below horizontal
	or		
	Three K-17-6 in.	Fixed	Trimetrogon

NOTE

On certain airplanes,[1] the forward camera bay is designated station 1 and the aft camera bay is designated station 2. In addition to the equipment listed above, the following items of equipment may be installed in these airplanes.

Location	*Equipment*	*Type Mount*	*Optical Axis*
Station 1	K-17C-6 in.	Fixed	Vertical
	K-17C-12 in.	Fixed	Vertical
	T-11	Fixed	Vertical
Station 2	CAS-2a	Fixed	25°, 35° or 90° below horizontal
	K-17C-6 in.	Fixed	Trimetrogon
	K-37 (modified)	Rotary	Vertical
	K-38-24 in.	Rotary	Vertical
	K-38-36 in.	Rotary	Vertical
	K-17C-6 in.	Rotary	3°, 15° or 90° below horizontal
	K-17C-12 in.	Rotary	3°, 15° or 90° below horizontal
	K-17C-6 in.	Fixed	Vertical
	K-17C-12 in.	Fixed	Vertical

A photo recorder, located on the forward nose bay deck, includes the following equipment:[2]

Radio Altimeter Repeater Indicator

Sensitive Altimeter

Airspeed Indicator

Flasher and Relay Unit

Remote Compass Repeater Indicator

Clock

Outside Air Temperature Indicator

Film Speed Indicator

Three Digit Subtractive-type Counter

Camera AN-N-6A (equipped with 17 mm lens)

[1] F9F-6P airplanes ser No. 134446 and subsequent.

[2] F9F-6P airplanes ser No. 127473 and subsequent.

The following additional equipment is installed:[1]

Description	For	Location
CK-1 Control Unit	S-7-S Sonne Camera	Cabin
Amplifier		Fuselage
Interval Control Computer Unit	K and CA-8 Cameras	Cabin
Film Breakage Warning Lights		Cabin
Three-digit Subtractive Type Counter		Cabin
Individual Camera Selector Switches		Cabin
Main Camera Power Switch		Cabin
Camera Position Selector Switch for Rotary Mount, and Position Lights		Cabin
Jet Air Ejector		Fuselage
Amplifier for Multi-camera Interval Controller		Fuselage
Camera Compartment Temperature Indicator	All Cameras	Cabin
Viewfinder A-10VF	All Cameras	Cabin

The following additional equipment is installed on certain F9F-6P airplanes.[2]

Description	For	Location
Master Control Unit CCS-1	All Cameras	Cabin
Photo Control Unit STA 1	All Cameras	Cabin
Photo Control Unit STA 2	All Cameras	Cabin
Flare Ejector Control Panel	K-37	Cabin
Ejector Pods	K-37	Bomb Racks
A-6, B-4 Flash Cartridge Ejector	K-37	Bomb Racks
Scanner	CAS-2a, K-17C	Fuselage
Scanner Converter	CAS-2a, K-17C	Fuselage
Servo Power Unit	CAS-2a, K-37, K-17C	Fuselage
Photo Observer		Cabin
Shutter Trip Control	K-37	Fuselage
Overrun Control	K-37	Fuselage
Film Failure Lights	All Cameras	Cabin
Operation Indicator Lights	All Cameras	Cabin
Oblique Sights	CAS-2a, K-37, K-17C	Cabin
Computer	CAS-2a	Cabin

NIGHT PHOTOGRAPHY INSTALLATION.[2]

The night photography installation incorporates a K-37 camera with an MA-10a magazine and 12 inch lens cone at a 90 degree angle in the station 2 camera bay. The shutter trip control for the K-37 camera is installed in the station 1 camera bay. The MA-10a magazine is operated by the servo power unit set for a 90 degree angle and 12 inch focal length. The interval controller, set for a 90 degree angle and 12 inch focal length, and with the altitude and ground speed control wheels on the MASTER control panel arbitrarily set to provide pulse to the flare ejectors, will operate the A-6 and/or B-4 flash cartridge ejectors.

BALLAST.

Approximately 100 pounds of ballast is installed between sta 12 and 25. This ballast is considered fixed equipment and is included in the empty weight of the airplane. It assures that the cg of the airplane will stay within the proper range when the heavy camera installations are used. An additional 120 pounds (approximately) of ballast is included under useful load. This additional ballast is installed when the lighter camera installations are used and is removed when the heavier installations are used. The 120 pound ballast is bolted underneath the fixed ballast.

Note

All ballast must be installed except when trimetrogen camera installation is used in combination with forward bay installation.

CAMERA COMPARTMENT AIR CONDITIONING SYSTEM.

Engine bleed air is ducted to fan shaped outlet nozzles installed inside of and forward of each of the camera windows, the scanner window and the viewfinder window. This air serves the dual purpose of supplying heat to the camera compartment to permit cold weather and altitude operation of the equipment and also defrosting of the windows. Temperature in the compartment is maintained at approximately 32.3°C (90°F) by an electronic temperature regulator and a temperature pickup. A thermometer records camera compartment temperatures on an indicator on the left console. Heat can be controlled by the adjacent switch.

CAMERA COMPARTMENT HEAT CONTROL.

This is an ON-OFF switch located on the wing flaps control panel outboard of the rudder trim tab control. It functions through relays in the camera compartment heating circuit to supply conditioned air for heating and defrosting the camera compartment and camera windows.

INDICATORS.

The outside air temperature and camera compartment temperature indicators are located immediately aft of the wing flaps control panel on the left console. The outside air temperature indication is received from a thermometer bulb mounted on the left side of the nose section, and the camera compartment temperature from a bulb mounted inside the camera compartment.

CAMERA WINDOW WASHING SYSTEM.

The camera windows are automatically washed at the beginning of each flight to remove salt spray or dust

[1] F9F-6P airplanes ser No. 127473 through 131255.

[2] F9F-6P airplanes ser No. 134446 and subsequent.

that might distort an image seen through the windows. A one gallon wash fluid tank is installed on a shelf on the left side, immediately forward of the fuselage nose section door and step assembly. When the airplane becomes airborne, a landing gear nutcracker switch operates a solenoid valve in the camera compartment air conditioning system which permits air at 10 psi pressure to feed into the tank and force water through the system to the window outlets. The water outlet for each window consists of a manifold of 1/16 inch tubing inserted through slots in the leading edge of each window frame to spray the water back and over the outside of each window. Low water level in the tank opens a float valve switch and shuts off the solenoid valve. A tank drain line is routed through the shelf and forward to drain overboard through the bottom aft camera window. Access for draining is through the open window.

Note

Use mixture consisting of 40% Methanol, Specification MIL-M-6999, and 60% water.

CAMERA WINDOW COVERS.

To protect the side windows from damage while the airplane is being serviced on the ground, covers are provided for the left and right windows. Each cover is painted insignia red and is equipped with a three inch square by 1/4 inch thick rubber pad cemented to the inner side at the approximate center to contact the plate glass window, and a sponge rubber seal running completely around the edge to cushion the cover against the window frame. The upper edge of the cover has two pins which drop into two holes in the top of the window frame to support the cover, and the lower edge contains a bracket and thumb-screw, by which the cover is held securely against the window.

CAUTION

See that window covers are removed before flight.

FIGURE 4-8 DELETED

VIEWFINDER.

Note

Pending modification, viewfinders will not be installed by contractor prior to delivery of airplane. Ballast is installed to compensate for the weight of this item. A seal plate is mounted on the forward cabin bulkhead (in the place normally occupied by the viewfinder face) to maintain cabin pressurization.

The viewfinder is designed to furnish the pilot with a clear, unobstructed view of the terrain below and ahead of the airplane for line-of-flight navigation and for locating the photographic targets. The viewing system consists of two integrated optical systems, one covering a wide field at a small scale and the other including a smaller field at a larger scale for determining drift, ascertaining or checking camera intervals, angular camera coverage, and for pinpointing or accurately positioning the airplane over any desired reference point.

The wide angle objective system has an included field of 85 degrees, with the optical axis tilted forward and upward in the longitudinal plane, 34 degrees from the vertical. This system furnishes the pilot with an extended view from near the forward horizon to a point five degrees behind the vertical, or nadir point. Suitable grids having controlled illumination are provided in both systems for the solution of drift problems, location of track line and the determination of the field of the various cameras.

The drift system has an included field of 30 degrees and the optical axis is vertical. This system covers an area directly below the airplane, at a larger scale, for navigation purposes, including the determination of drift. It also serves as a photographic viewfinder for cameras having relatively small field angles.

The second images as formed in the eyepiece and viewed by the observer, have a clear diameter of six inches and are direct and unreversed. Both images are formed selectively in the same plane, located in the common eye lens, by means of a changeover mirror system actuated by a servo-mechanism. The double exit pupil of the system provides full binocular vision with a lateral movement of the pilot's head of one and one-half inches for an interpupillary distance of 65 mm. The objective lenses of the two systems are so located that entrance pupils will pass through a single six inch diameter window.

A yellow filter, an opaque disc and a clear aperture are provided for selection at the discretion of the pilot.

CONTROLS.

All controls for the viewfinder are installed in the upper left corner of the main instrument panel.

ILLUMINATION CONTROL.

This control is located to the left of the viewfinder. The face of the control shows two concentric rings, labeled OUTER and INNER and indicates light on in the viewfinder for WIDE ANGLE and DRIFT settings.

GRID ROTATION SWITCH.

This switch is located to the left of the illumination control and is a two position momentary toggle switch placarded LEFT—RIGHT. It is used to rotate the grid in the viewfinder as required for computation of drift and ground speed by the pilot. It energizes a motor in the viewfinder body which rotates the grid through an arc of 30°—15° either side of zero reading, indicating direction of flight.

FILTER SWITCH.

This switch is located to the left of the grid rotation switch and is a three position selector switch marked A—B—C. It is used to select the yellow filter, opaque

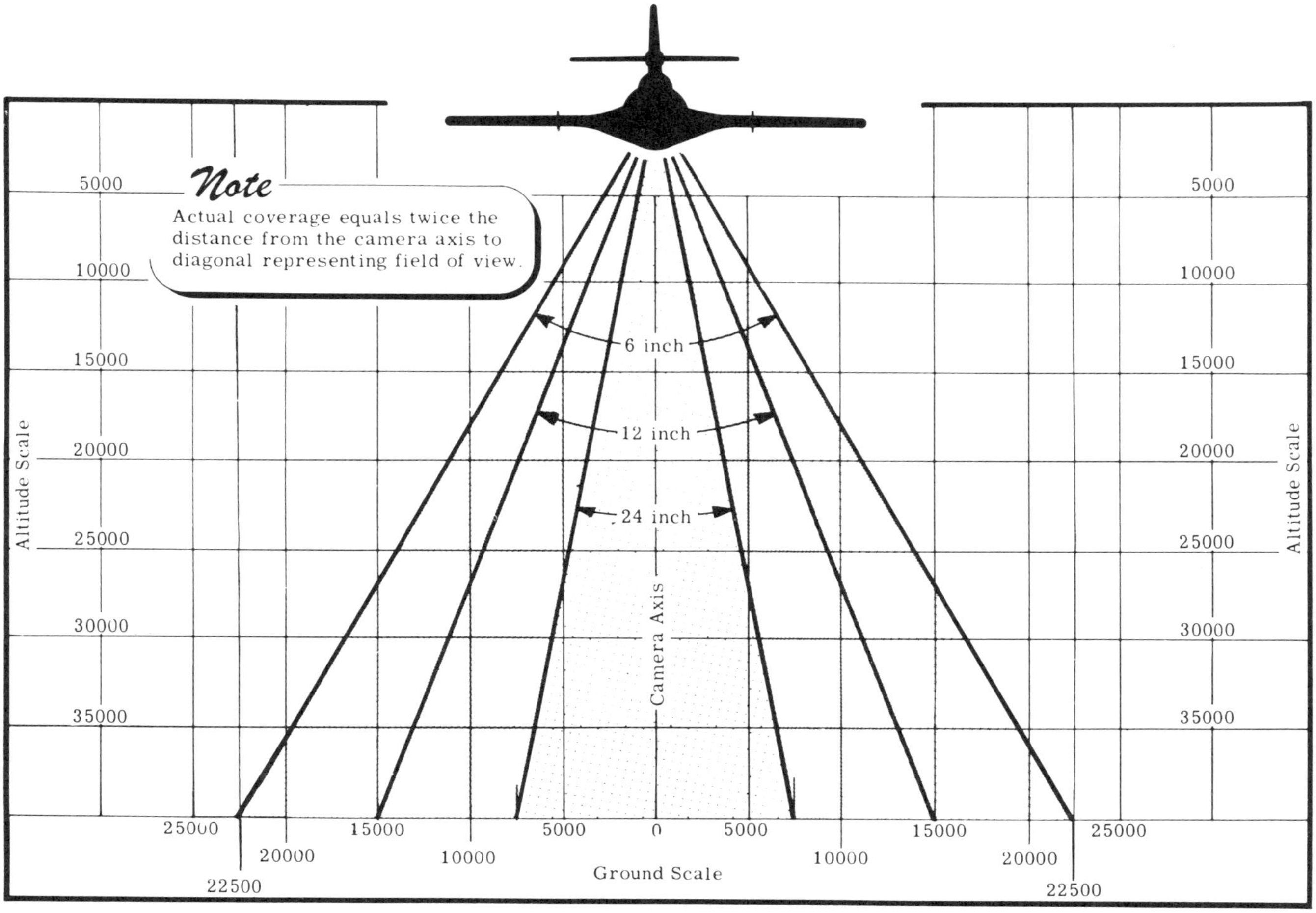

Figure 4-9. Vertical Camera Coverage Chart

disc, or clear vision. The opaque disc and filter are contained within the housing of the viewfinder and are moved into or out of the line of vision by an electric motor which operates when the switch is turned to the desired setting.

WIDE ANGLE—DRIFT SELECTOR SWITCH.

This switch is located immediately to the left of the filter and is used for setting the viewfinder, through a changeover mirror system, for the desired operation as the pilot first picks up and then pinpoints his photographic target, or for using the viewfinder as a driftmeter.

VIEWFINDER POWER SWITCH.

This is an ON-OFF switch located at the extreme left of the main instrument panel. This switch must be in the ON position to make the viewfinder operative.

On some F9F-6P airplanes,[1] the viewfinder power switch is located to the left of the viewfinder on the main instrument panel below the grid rotation switch and is marked VIEWFINDER PWR ON.

OPERATION.

When approaching a target, shift to the vertical drift system and remove the drift in exactly the same manner as with a conventional driftmeter. Shift back to the wide-angle system, which will show the entire area ahead of the airplane. Note that the track line in the wide-angle grid has moved, relative to the airplane heading, in the same direction as the drift correction. If the drift has been fully corrected, the wide-angle track line will now define the true track of the airplane over the ground, and any necessary adjustment in the airplane heading must be made to bring the track line into coincidence with the desired line of flight over the ground. After bringing the airplane to the correct position over the line of flight, it must be held there by frequent drift and heading corrections. Under certain extreme conditions of lighting, it will be necessary to introduce the glass viewing filter.

DRIFT CORRECTION.

The proper procedure for drift correction is essentially the same as that used with the B-3 driftmeter. Proceed as follows:

[1] F9F-6P airplanes ser No. 134446 and subsequent.

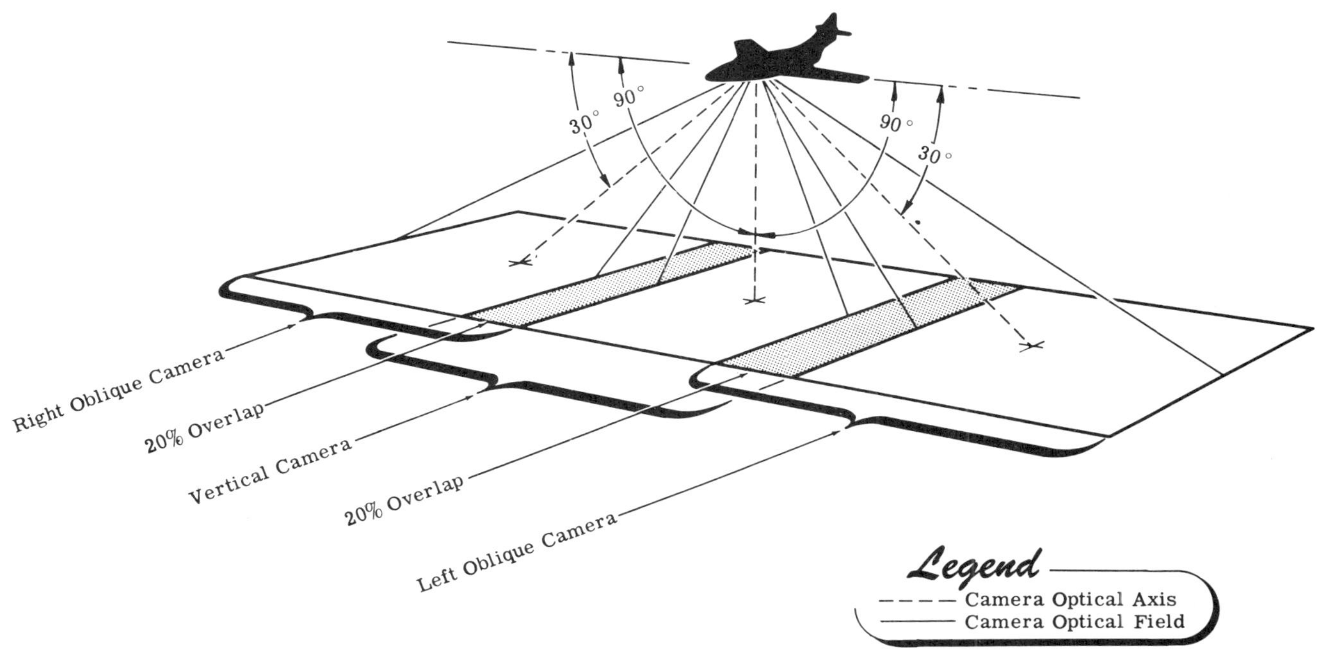

Figure 4-10. Trimetrogon Camera Coverage Chart

a. Choose a point of detail in the image which falls on or near one of the tracking lines parallel to the center track line, and follow the point back across the field of the instrument.

b. Make the necessary change in the drift angle of the grid to bring the track of the image point parallel to the grid lines.

c. When fully corrected, all points in the image will track parallel to the grid track lines.

d. Bear in mind that, after correcting for drift, any change in the heading of the airplane to bring the wide-angle track line into alignment with the flight line may introduce some drift error and require another drift correction.

DETERMINATION OF GROUND SPEED.

Determination of the true ground speed of the airplane follows the method and procedure used with other driftmeters. This procedure consists of the two following steps:

a. Timing the travel of an image point across a fixed distance on the grid with a stop watch.

b. Multiplying the time in seconds by a constant for the instrument. The second step is performed with the circular slide rule, supplied with the viewfinder and mounted on the right canopy rail. The image point is timed across the square drift grid, giving "t" in seconds. On the circular slide rule, set seconds on the inner scale, opposite the altitude on the outer scale, and read the true ground speed opposite the appropriate index for knots or mph. Because the optical system of the driftmeter contains no stabilization, it is essential that the airplane be maintained stable in a normal flight attitude.

CAMERA CONTROLS AND INDICATORS.[1]

All camera controls and indicators are in the pilot's cabin. The left console, just outboard of the throttle quadrant, is divided into three panels placarded FWD. BAY, AFT BAY and STRIP CAMERA.

FORWARD BAY PANEL.

The FWD. BAY panel contains the following items:

a. An ON-OFF switch identified K & CA CAMERA POWER, for starting camera operation.

b. An indicator light marked film feed, which glows when the film breaks in the K or CA-8 camera installed in the forward camera bay.

c. An AUTO-OFF-MANUAL switch, which is operated after the K & CA camera power switch to select either automatic camera operation through the interval controller or manual operation through the stick trigger switch.

d. A two-window indicator panel marked EXPOSURES REMAINING and FOCAL LENGTH, which shows the exposures remaining and focal length of the lens in use. The former indicator is operated electrically from the camera case; the latter is a placard marked by the ground crew.

AFT BAY PANEL.

The AFT BAY panel contains the following items:

[1] F9F-6P airplanes ser No. 127473 through 131255.

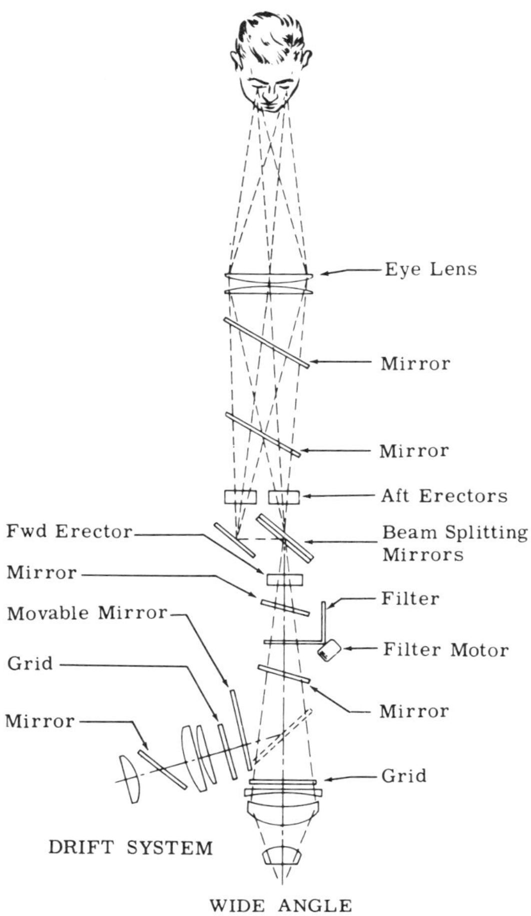

Figure 4-11. Viewfinder Optical System

a. A rotary camera position selector switch marked LEFT, 3°, 15°—VERTICAL—RIGHT, 3°, 15°. This switch supplies power to the rotary actuator on the camera mount when the rotatable K-17-24 in. or K-18-24 in. cameras are installed. Limit switches, mounted in the camera bay, and a cam on the actuator body serve to stop the rotating camera at the desired position by cutting the power supply at the position set by the selector switch. Each position indication over this selector contains a small electric bulb which glows when the check button is pressed if the camera has assumed the selected position.

b. A push button marked CAMERA POSITION CHECK BUTTON, located immediately inboard of the camera position selector switch handle. When pressed, this button will cause the light which corresponds to the position the camera is actually set in, to glow. This provides a double check for the pilot in the event mechanical failure has caused rotation to overrun or underrun the selected position.

c. Three film feed indicator lights placarded LEFT, VERT and RIGHT, which glow when the film breaks in the respective camera or cameras installed in the aft bay.

d. An AUTO-OFF-MANUAL switch, which is operated after the K & CA camera power switch to determine automatic camera operation through the interval controller, or manual operation through the pilot's stick trigger switch.

e. A two-window indicator panel marked EXPOSURES REMAINING and FOCAL LENGTH, which shows the exposures remaining and focal length of the lens in use. The former indicator is operated electrically from the camera case; the latter is a placard marked by the ground crew.

STRIP CAMERA PANEL.

The STRIP CAMERA panel contains the following items:

a. A power ON-OFF toggle switch, which is the main power supply control for the S-7-S Sonne continuous strip camera, when installed, and for the servo transmitter in the CK-1 remote control circuit.

b. A momentary FAST-SLOW film speed control switch, which controls the speed of film travel over the exposure slit in the continuous strip camera. It is also used for ground testing and adjusting film speed and slit width and for setting film speed as near as possible to that required for the mission. The FAST-SLOW switch is also used when the pilot desires to operate the camera manually.

c. An AUTO-OFF-MANUAL switch, which is operated to start the automatic control amplifier of the CK-1 remote control unit and supply power to the scanner, when in the AUTO position. The trigger switch on the control stick must be used to operate the camera for both manual and automatic operation. In automatic operation, the film speed is automatically adjusted by the scanner and amplifier to that of the image. In manual operation, the film speed will be as set by the FAST-SLOW switch.

d. A three-digit subtractive counter indicator, parallel with a subtractive counter on the camera case, indicates the number of feet of unexposed film remaining in the strip camera.

MAIN INSTRUMENT PANEL CAMERA CONTROLS AND INDICATORS.

The main instrument panel contains camera controls and indicators mounted horizontally across the top of the panel as follows:

a. The camera operating indicator lights are located immediately to the left of the viewfinder. They are placarded CAMERA OPER., and individually designated FWD. BAY and AFT BAY. These lights glow when the shutters are tripped on the respective units. Both are of the press-to-test type.

b. The film speed indicator is located to the right of the viewfinder and indicates the rate of film travel through the S-7-S camera. It is calibrated to read in inches per second and is a remote control, selsyn operated instrument. This indicator has three openings in the face and a turn knob in the center. The turn knob permits either the left window to open and the right to close when the 4-1/2—6 reading is set in the top center window, or the left window to close and the right to open when the 3—4 reading is set in the top window. Under either condition, film speed is read in inches per second through the left or right window, according to the lens cone in use at the time. The 4-1/2—6 scale is used with the 6 inch cone and the 3—4 scale with the 100 mm lens cone.

c. The film feed light is located slightly to the right of, and below the film speed indicator. This light glows green to indicate that film is feeding through the S-7-S camera and, when out, indicates that the roll has been expended or the film has broken or jammed.

d. Two synchronization lights, one red and one green, placarded SYNC, are located above the film speed indicator and between it and the viewfinder face. They indicate synchronization between the scanner, which follows airplane speed, and the S-7-S camera film speed. Lack of synchronization is corrected automatically by the scanner pulsing relays adjusting the camera speed to meet airplane speed, as interpreted by the scanner. This adjustment is slow, however, and the indication is given to the pilot so that he can adjust airplane speed to assist in achieving synchronization earlier.

CONTROL STICK TRIGGER SWITCH.

The trigger switch (usually used to fire guns) is wired to control camera operation through the MANUAL position of the switches on each of the three control panels on the left console and the AUTO position of the strip camera switch. Wiring also provides for interrupting the automatic operation for taking extra pictures when required. With the Sonne strip camera in use, control is only through the control stick trigger switch.

CAMERAS INSTALLED INDICATOR.

This indicator is located on the left cabin rail and is designed to receive labeled strips which serve to notify the pilot what cameras are installed in the camera bays.

CAMERA CONTROLS AND INDICATORS.[1]

There are three camera control panels on the left console outboard of the throttle quadrant marked MASTER, STA 1 and STA 2. An alternate power switch, marked CAM REMOTE MASTER, is located on the left cabin rail. The FLARE EJECTOR switch is located on the right console. Additional camera operating controls and indicator lights are mounted on the main instrument panel.

STA 1 CONTROL PANEL.

The STA 1 panel contains the following items:

a. At the lower right of the panel, a switch marked MASTER, OFF, and STICK, which is used to determine the operating mode selection for the station 1 camera bay.

b. At the top right of the panel, an indicator light marked READY. When the light is on, it indicates that the station is electrically connected and that power is available for operation.

c. An indicator light marked FILM FAILURE, located on the left of the panel, which is out under normal operation and on when difficulty develops.

d. A window indicator counter, marked EXP OR FT REMAINING, located in the center of the STA 1 panel, operates from a pulse originating within the equipment of the MA-10a, the T-11, or the CAS-2a installations. Magazines not so designed have the counters operating from a trip pulse to the camera.

MASTER CONTROL PANEL.

The MASTER control panel contains the following items:

a. At the lower right of the panel, the power ON-OFF switch. When this switch is set to ON, the READY indicator lights on the STA 1 and STA 2 panels will go on, indicating that the stations are in readiness for operation.

b. A three position, centrally located scanner-computer switch on the MASTER control panel provides the required information necessary to control image velocity. The scanner system operates only over the low altitude range, and scanner control is accomplished by the left position, marked SCANNER. Image velocity, as determined with the use of the CAS-2a camera system or the MA-10a magazine and auxiliary components, requires manual setting in altitude and speed by means of the ALT and KTS control wheels at the center of the MASTER control unit. Speed range settings are from 150 to 700 knots. Changeover to manual is effected by setting the three position scanner-computer switch to the center LO position for low altitude ranges of 100 to 5000 feet and to the right HI position for high altitude ranges of 1000 to 50000 feet.

c. An ON-OFF operate switch on the lower left of the MASTER control panel. Setting this switch to ON will initiate operation of all cameras set up for continuous or master operation.

STA 2 CONTROL PANEL.

The STA 2 control panel contains the following items:

a. A switch at the lower right of the panel marked MASTER, OFF, and STICK. This switch is used to provide the operating mode selection for the station 2 cameras.

b. At the top right of the panel, an indicator light marked READY. When the light is on, it indicates that the station is electrically connected and that power is available for operation.

c. A rotary camera mount position selector switch is located at the top center of the STA 2 panel. Positions are marked LEFT 3°, 15°—V—RIGHT 3°, 15°. This switch supplies power to the rotary actuator on the

[1] F9F-6P airplanes ser No. 134446 and subsequent.

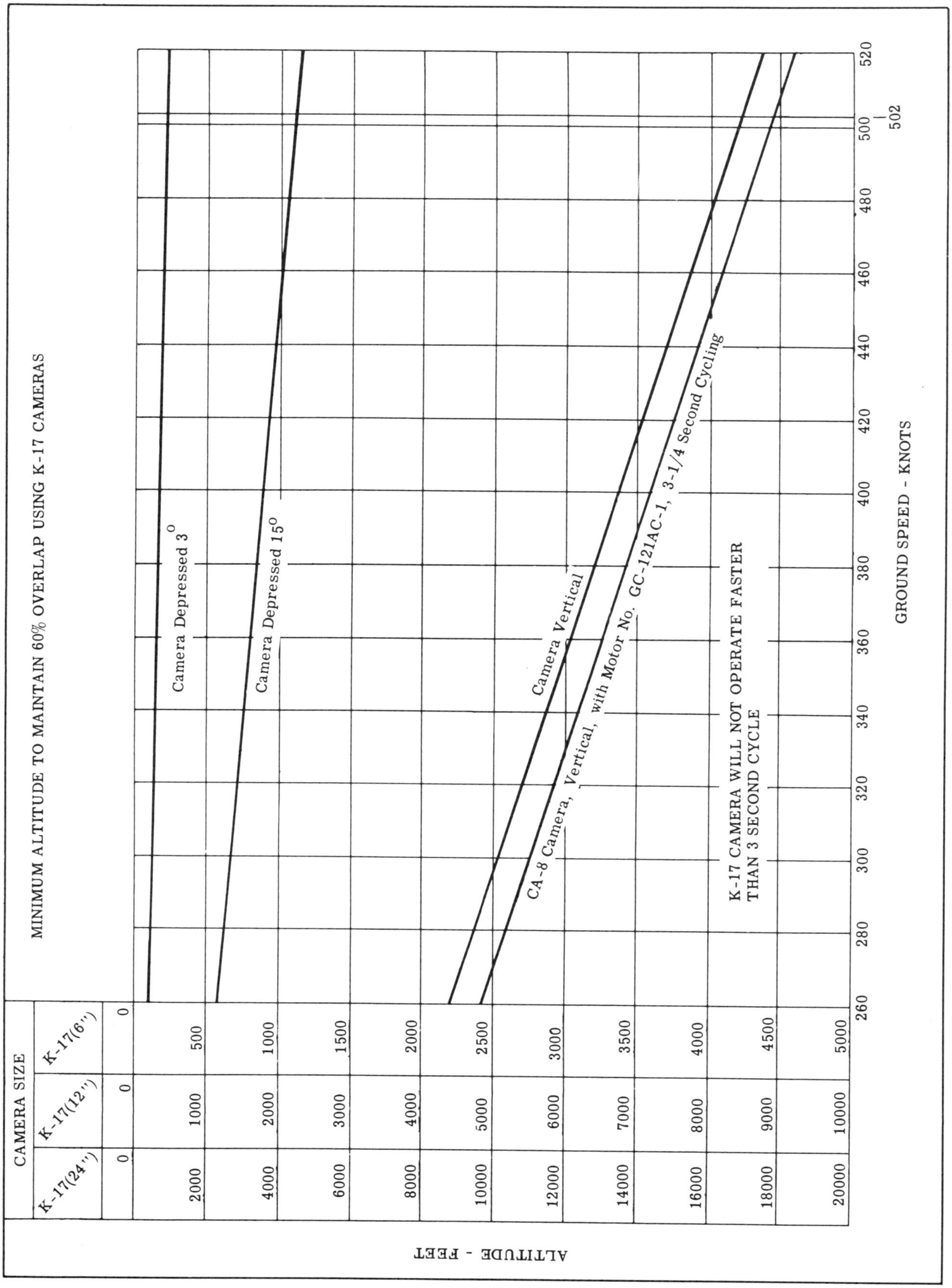

Figure 4-12. Minimum Altitude to Obtain 60% Overlap—K-17 Cameras

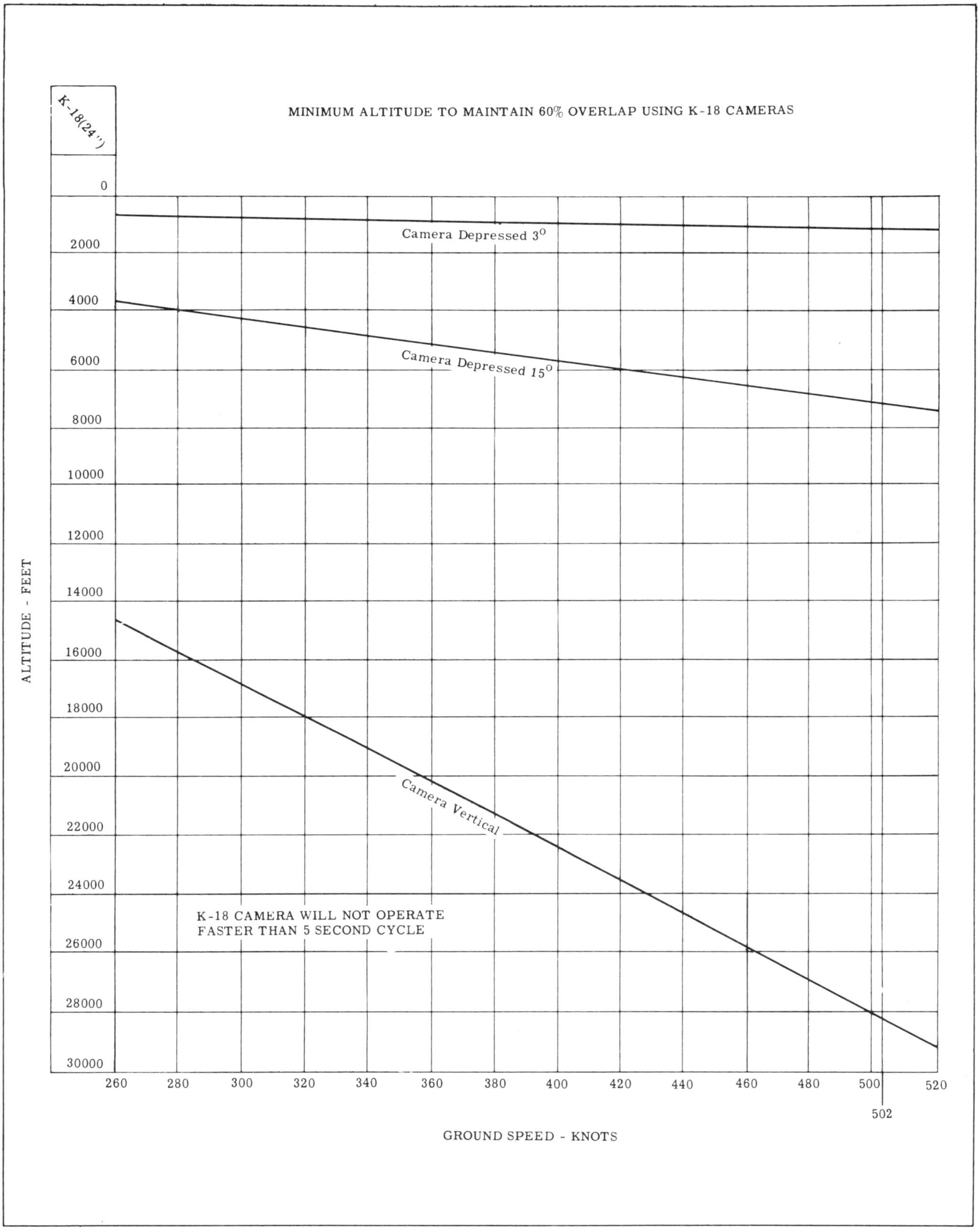

Figure 4-13. Minimum Altitude to Obtain 60% Overlap—K-18 Cameras

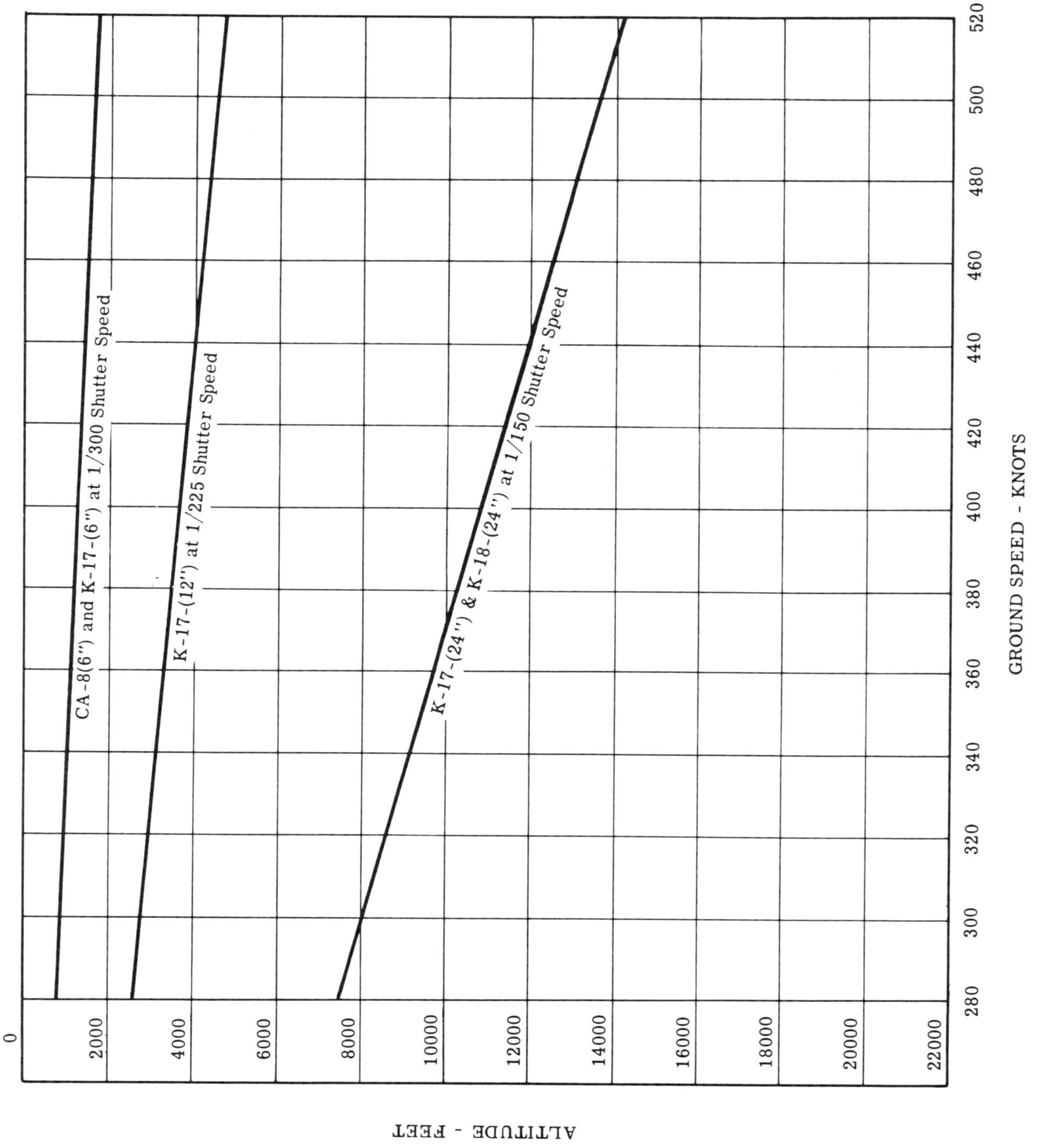

Figure 4-14. Minimum Altitude to Stop Motion at Maximum Shutter Speed

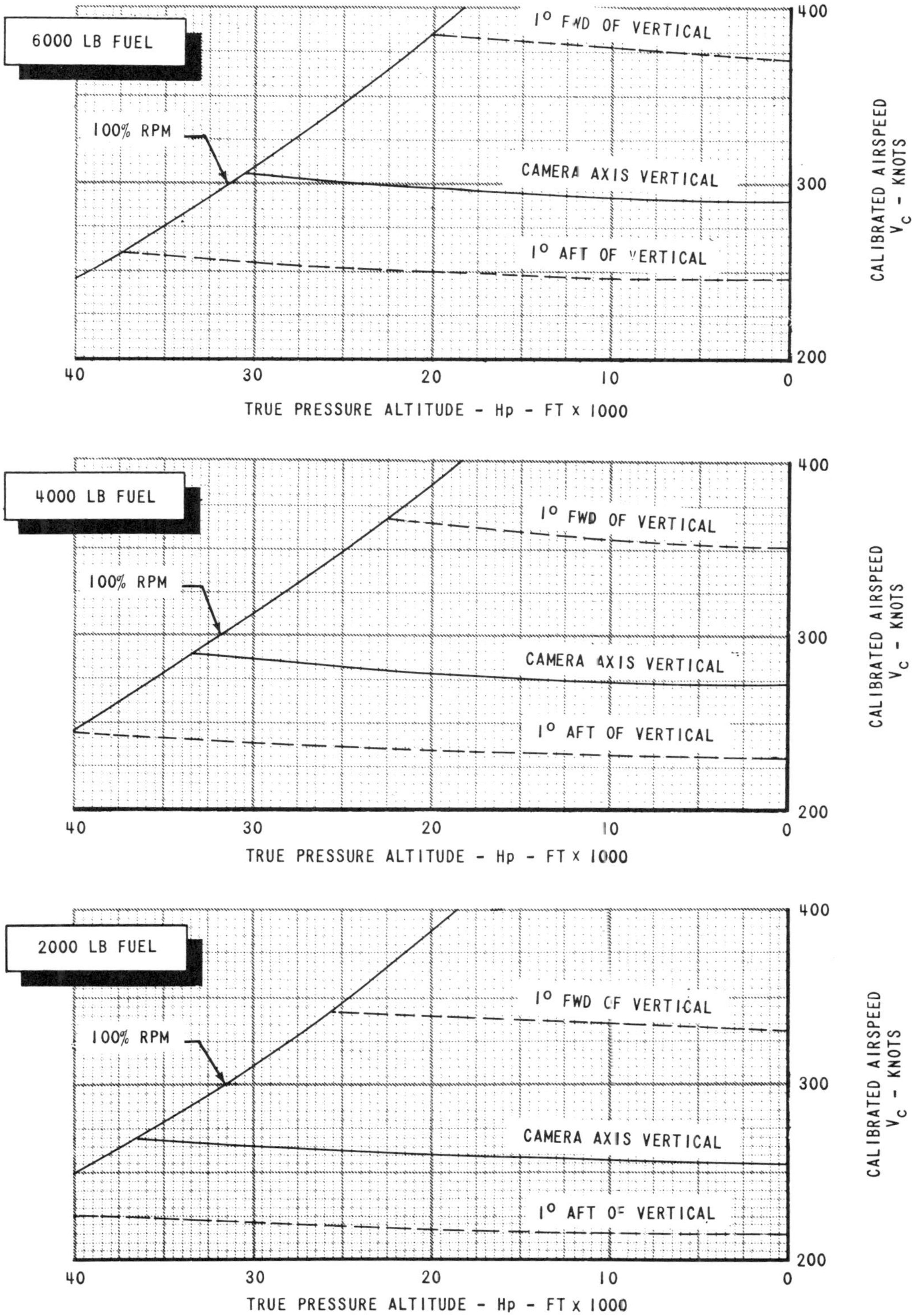

DATA BASIS: G.A.E.C. FT79-0-Id. 1

NOTE: GROSS WEIGHT MINUS FUEL - 11445 LB

Figure 4-15. Airspeed for Vertical Photography

camera mount when the rotatable K-17C cameras, the K-37 (modified) or the K-38 cameras are installed. Limit switches, mounted in the camera bay, and a cam on the rotary mount, serve to stop the mount at the desired position by cutting the power supply at the position set by the selector switch.

d. At the upper left, a light marked POSITION, which glows when the rotary camera mount, if in use, is in the incorrect position.

e. An indicator light marked FILM FAILURE, located on the left center of the panel, is out under normal operation and on when difficulty develops.

f. A window indicator counter, located in the center of the STA 2 panel, marked EXP OR FT REMAINING. It operates from a pulse originating within the equipment for the MA-10a magazine or the CAS-2a installations. Magazines not so designed have the counters operating from a trip pulse to the camera.

g. At the lower left corner of the panel, an indicating light, marked INADEQUATE LIGHT, which is provided for use with the CIL-1, -2, -3 or -4 automatic lens cone when installed on the CAS-2a camera. Normally out, this light glows when photographic brightness values fall too low to permit proper operation of the automatic aperture control system.

h. A rotary switch at the lower center of the panel provides exposure control for the CAS-2a camera installed in the station 2 camera bay. Positions are AUTO, for automatic control, and 1/50, 1/100, 1/200, and 1/400 of a second shutter speeds.

MAIN INSTRUMENT PANEL CAMERA CONTROLS AND INDICATORS.

The main instrument panel contains camera operation and film failure indicator lights, mounted near the top of the panel to the left and right of the viewfinder, as follows:

a. To the left of the viewfinder are camera operating lights provided as alternate lights. They are marked CAMERA OPER, and individually designated STA 1 and STA 2. They glow when the camera is pulsed.

b. The viewfinder power switch is located to the left of the camera operating lights.

c. Above this switch is the grid rotation switch, marked LEFT and RIGHT.

d. To the right of the viewfinder are alternate FILM FAILURE lights, marked STA 1 and STA 2. The lights are on only when difficulty develops.

REMOTE CAMERA CONTROLS.

A camera remote master ON-OFF switch located on the left cabin rail provides an alternate means of camera operation initiation. In addition, the control stick trigger switch (usually used to fire guns) is wired to control camera operations through the STICK position of the mode selector switches on the STA 1 and STA 2 panels on the left console.

NIGHT PHOTOGRAPHY CONTROLS.

The controls for setting and operating the K-37 camera in night photography are the same as indicated for the STA 2 control panel except that the following controls for flare ejection are included:

a. A flare release panel is provided on the right console. The flare ejector switch has three positions, marked LEFT, OFF and RIGHT. These positions select the photo flare ejector (A-6 or B-4), located in the left or right wing pods, to be used during night operation.

b. Camera operation may be initiated at the STA 2 panel by setting the mode selector switch to MASTER or STICK. When the flare ejector switch is moved to LEFT or RIGHT, the interval controller transmits pulses directly to the flare ejector for automatic release of the flare. The illumination provided by the flare, in turn, operates the shutter trip control. The overrun device provides a five second delay in release of magazine and/or camera operation as insurance against the loss of the last flare photograph.

c. With the controls set as in step b. above, operation of the extra picture button on the control stick provides the pulse for the release of one flare, starts the MA-10a magazine, and actuates the overrun device to ensure obtaining the exposure. An interlock circuit used only in night photography prevents operation of the extra picture switch while exposures are taken with the mode selector switch set at MASTER or STICK.

d. Two indicator lights, marked FAILURE, at the top left of the flare release panel are normally out except when failure occurs in the photo flare ejector or when the ejectors are empty.

e. Two guarded ON-OFF emergency switches at the center of the flare release panel, marked JETTISON and SALVO, are used for pod jettison and flare salvo functions. Interlocks prevent the flares from being released unless the landing gear control is in the UP position.

f. Initial operations for the release of the M-120 flash bomb is accomplished at the MASTER and STA 2 control panels. When making a photographic run at high altitudes, and the bomb is to be released, the flare ejector switch is placed in the RIGHT or LEFT position. The interval controller transmits pulses directly to the bomb releasing mechanism, dropping the bomb. The firing mechanism becomes operative and the illumination, in turn, operates the shutter trip control.

g. To jettison the M-120 flash bomb, operate the drop tank jettison switch located on the left console.

CAUTION

Avoid contact with the drop tank jettison switch when the bomb racks are loaded with the M-120 flash bombs. Interlocks are not provided to prevent the release of bombs when the landing gear is down.

MULTI-CAMERA INTERVAL CONTROL—TYPE MCD-2.

This equipment is composed of an electronic interval controller (timer and magnetic amplifier assembly) and the interval computer on the right console. These components initiate and control electrical pulses of variable intervals used to cycle the cameras. Manual setting of altitude or ground speed on the cabin control unit positions a logarithmic potentiometer which, through the magnetic amplifier, varies the output speed of a magnetically controlled variable speed motor of 3°, 15°, and 90°. Each of these cam groups turns at a different speed but at a fixed ratio to the others, corresponding to their depression angles. Contact springs riding on these cams close electrical circuits to cycle the cameras in proper sequence to provide 60 per cent forward overlap. This control unit is designed to operate correctly through an altitude range of 1000 to 55000 feet and at a speed range of 280 to 502 knots. Thermostatically controlled electrical heaters are provided to ensure operable temperatures within the amplifier and timer enclosures when the ambient temperature is as low as —34.5°C (—30°F.) This equipment requires both a-c and d-c power supply. The load is 385 volt-amperes continuous at 115 volts, 400 cycle, and 3/4 ampere at 27-1/2 volts dc. It is not necessary to change the altitude and speed setting when taking oblique photographs to obtain 60 per cent forward overlap. In all instances, when the MCD-2 interval control is operating and properly set for ground speed and altitude above terrain, the cameras will be triggered at the proper intervals to obtain 60 per cent forward overlap on oblique cameras as well as vertical cameras. It is not necessary to calculate the slant distance when taking oblique photographs. The rotatable cameras switch automatically modifies the signal in the MCD-2 to always provide 60 per cent overlap at the center of the photograph regardless ofthe camera angle . When the MCD-2 is not functioning, or when it is desired to take overlapping photographs manually, obtain the interval for vertical photographs by using the viewfinder. For calculating the interval for oblique photographs, see figure 4-8.

INTERVAL COMPUTER.

The interval computer of the multi-camera interval control is mounted in the right console and consists of two wheel controls, mounted vertically and extending slightly out of the panel, on which the pilot sets true ground speed and altitude. The wheels are identified ALTITUDE (X1000) and GROUND SPEED (KNOTS). These figures are transmitted to the timer and magnetic amplifier assembly, located on the right side of the ammunition shelf in the nose wheel well. This assembly is an electronic device which uses these figures to establish a pattern of electrical impulses for actuation of the K or CA-8 cameras to provide a 60 per cent forward overlap of photographs.

OPERATION—MULTI-CAMERA INTERVAL CONTROL.

Set altitude (above terrain) and speed (ground speed) on the MCD-2 control on the right console. Turn K & CA camera power switch to ON. Set appropriate switch on left console to AUTO for cameras to be used. As soon as this switch is turned on, all cameras selected for automatic operation will operate automatically at intervals to obtain a 60 per cent forward overlap regardless of the angle of obliquity. Extra pictures between the automatically timed pictures may be taken at the option of the pilot by actuating the trigger switch on the stick. These extra pictures will not interrupt the automatic interval unless the intervals are less than double the cycling time of the cameras. At any rate, the regular interval will be reestablished after the extra pictures are taken when it is desired to ensure that some specific object is photographed directly in the center of the picture, or when the airplane is tilted to right or left to an attitude considered excessive for making a mosaic or for stereo pairs.

ECD-2 INTERVAL CONTROLLER.[1]

The ECD-2 electronic interval controller mounted in the nose wheel well computes and supplies accurately controlled electrical pulses and a precision reference d-c voltage to the camera control system. The unit also performs intervalometer functions for non-image-motion-compensating magazines in day photography operations and intervalometer functions for proper photo-cartridge (flare) release in night photography operations.

PREFLIGHT OPERATING INSTRUCTIONS.

The following procedure is performed by the ground crew.

Note

The STATION 1 station interval computer (on left, as viewed facing aft) controls the station 1 camera bay and the STATION 3 (right) station interval computer controls the station 2 camera bay. The STATION 2 (center) station interval computer of the ECD-2 interval controller is not used on F9F-6P airplanes.

a. Set the focal length switches on the ECD-2 unit to the proper effective focal length for the lenses being used in the corresponding camera stations.

b. Set the depression angle switch on the STATION 1 station interval computer to VERTICAL.

CAUTION

If the depression angle switch on the STATION 1 station interval computer is set to the MOUNT position, a camera "runaway" condition (continuous recycling) will result.

[1]F9F-6 airplanes ser No. 134446 and subsequent.

c. Set the depression angle switch on the STATION 3 station interval computer to MOUNT. This setting automatically transfers the depression angle selection function to the mount selector switch on the photo control panel on the left console.

d. Set the rotary mount selector switch at the top center of the STA 2 control panel (left console) to the VERTICAL position for all camera installations in the station 2 camera bay except the rotary mount installation when used at depression angles other than 90°, or the CAS-2a installation.

e. With the camera control system in operation, check the FUSE OUT indicator light on the ECD-2 controller. If illuminated, remove unit for bench check.

CAMERAS.[1]

Note

The following cameras may also be installed in later airplanes.[2]

K-18-24 IN. CAMERA.

The K-18 airplane camera is designed for making high altitude, large scale, vertical and spotting photographs and is installed in the aft bay. It accommodates a type A-7 or A-8 roll film magazine which produces a 9 by 18 inch negative. The lens cone and shutter assembly has a focal length of 24 inches. The lens cone contains the drive, the control mechanism and the electrical connections. The interval control computer is used for auto matic operation. The A-10VF vertical viewfinder is used in conjunction with the operation of the K-18 camera. Detachable lens filter units are provided for the lens cone.

OPERATING INSTRUCTIONS.

Controls for the K-18 camera are located on the left console, the pilot's control stick, and on the interval computer on the right console. The power is controlled by the K & CA camera power switch. The AFT BAY panel contains the camera position selector, the film feed lights, the AUTO-OFF-MANUAL control switch, and the exposures remaining and focal length indicators. The interval computer controls camera cycling to take a series of exposures having a 60 per cent forward overlap when both altitude and ground speed are set by the pilot. The trigger switch on the control stick is operated to secure extra exposures when using the interval computer or to operate the K-18 camera when manual control is desired.

CA-8 CAMERA.

The type CA-8 airplane precision mapping camera is an automatic camera designed for making a series of aerial photographs for mapping and charting purposes. It mounts in the forward bay. It is equipped with a six inch f:6.3 metrogon lens and a Navy type MA-6 magazine which gives approximately 250 nine by nine inch negatives on a 200 foot film roll. The interval controller and the viewfinder are used in conjunction with the operation of this camera.

The shutter speeds and diaphragm openings are as follows:

Speeds: 1/100, 1/200, 1/300 second.

Diaphragm: f:6.3, 8, 11, 16, 22.

Automatic operation permits the interval computer to send timing impulses to operate the camera shutter and advance the film to the next frame position to obtain 60 per cent forward overlap by compensating for airspeed and altitude.

The magazine is equipped with a vacuum back, around which the film must be routed, to ensure that the film is held flat in the focal plane. A jet air ejector pump supplies this power. The magazine is provided with a connection for an external indicator light, located on the main instrument panel. A blinking light indicates that the camera is operating.

OPERATING INSTRUCTIONS.

Controls for the CA-8 camera are located on the left console and the control stick. The interval control computer is located on the right console.

The power is controlled by the K & CA camera power switch. Immediately outboard of this switch on the forward bay panel is the film feed indicator light, the AUTO-OFF-MANUAL control switch, and the combination indicator which shows exposures remaining and focal length. When the three position switch is set to AUTO, it permits the interval control computer to send timing impulses to operate the camera shutter and advance the film to the next frame position to obtain 60 per cent forward overlap by compensating for airspeed and altitude.

The pilot's control stick trigger switch is used for manual operation or to by-pass the interval controller for the purpose of taking extra pictures when required.

K-17 AIRCRAFT CAMERAS.

The K-17 camera is designed to take vertical and spotting photographs. The camera accommodates a type A-5A or A-9 roll film magazine which gives a nine by nine inch negative. Three interchangeable lens cones and shutters, having focal lengths of 6, 12 and 24 inches, are available. Detachable photographic filter units are provided for the 12 and 24 inch lens cones. The K-17 with 24 inch lens cone and shutter is used in the aft bay and installed in the rotary mount. The K-17 with either the 12 or 6 inch lens cone and shutter may be installed in mounts in the forward bay for vertical photography. The mount supplied for installation of the K-17 with 6 inch lens cone and shutter will also accommodate the CA-8 camera for installation in the forward bay. The K-17 camera is used in the aft bay in the trimetrogon installation, which uses three K-17 cameras with 6 inch

[1] F9F-6P airplanes ser No. 127473 through 131255.

[2] F9F-6P airplanes ser No. 134446 and subsequent.

lens cones and shutters. In this installation, one camera is mounted with its optical axis vertical and tilted two and one-half degrees aft of a plane vertical to the fuselage reference line, and one each mounted on the left and right sides with their optical axes 60 degrees above the optical axis of the vertical camera. This results in a horizon-to-horizon athwart flight line photograph, with a 20 per cent overlap between negatives, which is particularly suitable for terrestrial mapping (figure 4-10).

OPERATING INSTRUCTIONS.

FORWARD BAY INSTALLATION.

When the K-17-6 in. or K-17-12 in. camera is installed in the forward camera bay, it is a fixed position camera, taking vertical photographs only. It is designed for use in conjunction with the viewfinder. Controls for the K-17-6 in. and K-17-12 in. cameras are on the forward bay panel and on the control stick. The interval control computer is located on the right console. Operation of any camera installed in the forward bay is similar. For operation, refer to operation of the CA-8 camera.

AFT BAY INSTALLATION.

When the K-17-24 in. camera is used, it is installed in the aft camera bay in a rotary mount. When the K-17-6 in. cameras are used, three cameras are installed in the trimetrogon mount. Controls are on the aft bay panel and on the control stick. For operation, refer to operation of K-18 camera.

S-7-S CAMERA.

The type S-7-S Sonne continuous strip camera is mounted in the forward bay and is used for taking sharp vertical photographs at low altitudes and high speeds. It is a semi-automatic aircraft camera, using film 9-1/2-inches wide in lengths up to 200 feet. In place of a shutter, the camera has a slit which controls exposure. When the camera is operating, the film moves past this slit in the same direction as, and in exact sychronization with the relative movement of the image in the focal plane. This synchronization of image with film speed over the slit produces a motion stopping effect. The exposure is governed by the speed of the film and the width of the slit.

The complete installation consists of the following principal components:

S-7-S Camera

Scanner

Amplifier

CM-1 Photo Recording Unit

CK-1 Remote Control Unit

The heart of this installation is the scanner which measures electronically the speed at which the ground passes under the airplane. The amplifier is a computer which transforms information from the scanner into usable signals for the CK-1 remote control unit. In automatic operation, this control translates this information into altitude, airspeed, slit width and exposure. The servo transmitter of the CK-1 unit takes this data and modifies it for focal length of lens in use. The camera itself, receiving all this information, drives the film at the correct speed.

An electrically operated film footage counter is provided on the camera body to transmit to an indicator in the cabin the number of unexposed feet of film remaining on the spool.

A green pilot light on the main instrument panel glows when film is pulling through the camera. The light receives its power supply from the line switch on the camera body, which is actuated by the pressure of the focal roller.

OPERATING INSTRUCTIONS.

The controls for the S-7-S camera are on the STRIP CAMERA panel on the left console. Actuation of the camera is by the control stick trigger switch only. With probable speed, altitude and light conditions that will be found on the specific mission known, and required ground settings for scanner and slit control established and set before flight, it is merely necessary for the pilot to set the control switch to AUTO position, and the power switch ON, about five minutes before he intends to use the camera. This allows time for the tubes to heat up to operating temperature and for the scanner and camera to be brought into synchronization. At this point, it becomes the function of the automatic film speed control equipment to translate the relative speed of the ground and airplane into a pulsating electric current and use this frequency as a measure of control over the film speed in the camera. It does this in the following manner:

a. Measures the speed of the ground image.

b. Measures the speed of the film.

c. Compares these speeds and determines the required correction in film speed, if any.

d. Makes this required correction.

In performing these functions, the scanner reads and interprets ground speed through two photo electric cells, which pick up the pulsating light waves and translate them into sine wave voltage. The servo generator unit converts camera focal roller speed (ultimate film speed) into sine wave voltage. The amplifier compares these frequencies and dictates the corrective measures by correcting film speed. With the camera ready for operation, the pilot may start photographing by squeezing the control stick trigger switch, and stop photographing by releasing the trigger.

The red and green synchronization lights on the main instrument panel, flashing evenly and continuously from red to green, indicate that film and airplane speeds are synchronized. When the red light glows, it indicates that film speed is below image speed and is being increased by the servo unit. When the green light glows it indicates that film speed is above image speed and is being decreased by the servo unit.

To operate the S-7-S camera manually, the control switch must be set to MANUAL and the FAST-SLOW switch adjusted to obtain film speed, determined by the Model W computer, for the airplane's airspeed and altitude above terrain.

Note

The airplane must then be flown precisely at the predetermined altitude and speed. Manual operation will not be required unless there is malfunction of the automatic system.

CAMERAS.[1]

K-17C CAMERA.

The K-17C camera is designed as an all purpose, vertical and oblique reconnaissance camera. When equipped with the 6 inch metrogon lens cone, the camera is capable of faster operation than the K-17, thereby enabling charting operations at higher airspeeds. In this camera installation, the K-17C is used as an alternate to the K-37 (modified) camera installation in the station 2 camera bay. The camera employs two interchangeable lens cone and shutter assemblies having focal lengths of 6 and 12 inches. The camera accommodates the MA-10a moving film magazine with a 9-1/2 inch roll film.

OPERATING INSTRUCTIONS.

The K-17C camera is controlled from either the STA 1 or STA 2 control panel, depending on the camera installation. When the K-17C camera is installed in the station 1 camera bay, it is fixed at 90 degrees. The power is controlled at the MASTER control panel, located on the left console, or the remote camera control ON-OFF switch on the left cabin rail. When the power switch on the MASTER control panel, or the alternate switch on the left cabin rail, is set to ON and the mode selector switch on the STA 1 control panel is set to MASTER or STICK, the READY lights on the STA 1 and STA 2 control panels will glow. When ready for the photographic run, the operate switch on the MASTER control panel is set to ON to initiate master or continuous operation. Where three K-17C cameras are installed as a trimetrogon installation, the STA 2 controls are similarly employed, as indicated in the preceding description of the station 1 installation. When a K-17C camera is installed on the rotary mount in the station 2 camera bay, operation is the same as described above, except that the mount switch on the STA 2 control panel must be operated to position the mount to the vertical or oblique angle to be used. The POSITION light should be out, indicating that the cameras are in the correct position for the photographic run.

K-37 (MODIFIED) CAMERA.

The K-37 (modified) camera, equipped with the MA-10a magazine, is used specifically for taking night photographs at high altitudes in conjunction with the use of pyrotechnic bombs or cartridges and syncroflash equipment. The camera is installed in the station 2 camera bay, employing the same mount as the K-17C camera. The modification to the K-37 camera consists of adding the capping shutter actuating mechanism. When the MA-10a system recycles, the recapping shutter is automatically positioned to prevent images from reaching the film during transport. The MA-10a magazine employs a 9-1/2 inch roll film.

OPERATING INSTRUCTIONS.

Controls for the K-37 (modified) camera are on the STA 2 control panel and the pilot's control stick (camera trigger and extra picture switches). The power is controlled at the MASTER control panel, located on the left console, or the remote camera control ON-OFF switch on the left cabin rail. When the power switch on the MASTER control panel, or the alternate switch on the left cabin rail, is set to ON, and the mode selector switch on the STA 2 control panel is set to MASTER or STICK, the READY lights on the STA 1 and STA 2 control panels will glow. On the flare ejector panel, the flare ejector switch should be set at LEFT or RIGHT to select the A-6 or B-4 ejector to be used during the run. When ready for the photographic run, the operate switch on the MASTER control panel or the master camera remote control switch on the left cabin rail is moved to its ON position to initiate master or continuous operation, respectively. Operation of the extra picture switch on the control stick can be used to provide for the release of one flash bomb and the resulting extra picture. The interlock circuit, used only in night photography, prevents operation of the extra picture switch while exposures are being taken with the mode selector switch set to MASTER or STICK.

K-38 CAMERA.

The K-38 camera is used specifically for taking large-scale spotting and reconnaissance photographs. The camera is installed in the station 2 camera bay, employing a modification to the rotary mount for the camera installation. The K-38 camera utilizes two interchangeable lens cone and shutter assemblies having focal lengths of 24 and 36 inches. One inch blocks are provided for installation of the K-38 camera with the 24 inch lens cone. Additional 1-1/2 inch blocks for the mount are provided for installation of the K-38 camera with the 36 inch lens cone. When the A-8B magazine is used with the K-38 camera, a 410 foot length of roll film 9-1/2 inches wide is used to produce approximately 260 negatives 9 inches by 18 inches. An interval control is used with the camera during automatic operation to provide the necessary intervals between exposures.

OPERATING INSTRUCTIONS.

Controls for the K-38 camera are located on the STA 2 control panel and the pilot's control stick (camera trigger and extra picture switches). The power is controlled at the MASTER control panel or at the remote camera control ON-OFF switch on the left cabin rail. When

[1] F9F-6P airplanes ser No. 134446 and subsequent.

the power switch on the MASTER control panel, or the alternate switch on the left cabin rail, is set to ON and the mode selector switch on the STA 2 control panel is set to MASTER or STICK, the ready lights on the STA 1 and STA 2 control panels will glow. When ready for the photographic run, the operate switch on the MASTER control panel is set to ON to initiate master or continuous operation. The extra picture button switch on the stick can be operated to take extra pictures when desired.

CAS-2a CAMERA.

The CAS-2a camera is designed for continuous strip aerial photography by means of manual or automatic operation. Automatic operations are twofold, incorporating the features of both the scanner and CILS-1, -2, -3 or -4 automatic lens cone systems. Inasmuch as this is a shutterless type camera, equivalent "shutter speeds" (1/50, 1/100, 1/200 and 1/400) are attained by a combination of slit widths and film transport speed. The slit width is controlled manually by the exposure switch (labeled with the equivalent shutter speeds and located on the STA 2 control panel) and automatically when the scanner or automatic lens cone systems are utilized. Film transport speed is controlled by the settings of the altitude and ground speed control wheels on the MASTER control panel. The information furnished by the settings of these controls is further modified by the factors of focal length and depression angle, which have been manually set in at the servo power unit, and the film emulsion rating, as set in at the ASA exposure index control on the lens cone. Power is controlled by the power ON-OFF switch on the MASTER control panel or the alternate remote camera control ON-OFF switch on the left cabin rail. The mode of operation (whether the camera will operate from the OPERATE switch on the MASTER control panel or from the control stick trigger switch) is selected by means of MASTER-OFF-STICK switch located on the STA 2 control panel.

The CILS-1, -2, -3 or -4 lens cone used with this camera includes a light sensing element and a servo system which automatically adjusts the diaphragm aperture and slit width to assure properly exposed negatives over an extremely broad range of light conditions.

The CAS-2a camera can also utilize standard stereo lens cones of 100mm, 7, 12, or 20 inch focal length.

OPERATING INSTRUCTIONS.

Controls for the CAS-2a camera are located on the MASTER and STA 2 control panels on the left console, plus the camera trigger switch on the pilot's control stick. For all operations, set POWER switch, or the alternate remote camera control switch on the left cabin rail, to ON at least five minutes before operating the camera.

For manual operation, proceed as follows:

a. Set exposure switch to anticipated exposure time (equivalent shutter speed), taking into account target brightness and emulsion sensitivity.

b. Turn camera ON and run a few feet of film through camera to ensure that camera is working properly.

c. Set KTS and ALT control wheels to correspond with existing flight conditions.

d. Set mode selection switch to desired mode of operation.

e. To take photographs, set operate switch to ON. If mode selector switch is set at MASTER, camera will start operating. If mode selector switch is set at STICK, press camera trigger switch to take photographs.

Note

If the film failure light glows, it indicates that film is not being fed through the camera properly.

For scanner operation, proceed as follows:

a. Set scanner-computer switch to SCANNER. The camera is then controlled automatically over the low altitude range only.

b. Follow steps b. through e. given under manual operation above.

To switch over from automatic to manual, set scanner-computer switch to center LO position for 100 to 5000 foot altitudes and to HI position for 1000 to 50000 foot altitudes.

For automatic lens cone operation, set exposure switch to AUTO. Then proceed with steps b. through e. as given under manual operation above.

Note

If light is insufficient, the INADEQUATE LIGHT indicator will glow steadily. The slit will then remain in a position equivalent to 1/25 second and the film should be stopped or camera operated manually. Sporadic, short blinks of this indicator light can be ignored, as they signify only that the lens cone is cycling to another shutter speed.

At conclusion of photographic run, turn camera off by setting operate and power switches to OFF.

T-11 CAMERA.

The type T-11 camera is used to take precise aerial photographs for the preparation of topographic maps, with pertinent data recorded on each exposure. This mapping camera is intended for full automatic operation and functions with a six inch, f:6.3, class T metrogon lens. The film roll is 390 feet long, permitting approximately 450 to 460 negatives 9-1/2 by 10-1/4 inches.

OPERATING INSTRUCTIONS.

Controls for the T-11 camera are located on the STA 1 control panel and the pilot's control stick (camera trigger switch). The power is controlled at the MASTER control panel or at the remote camera control ON-OFF

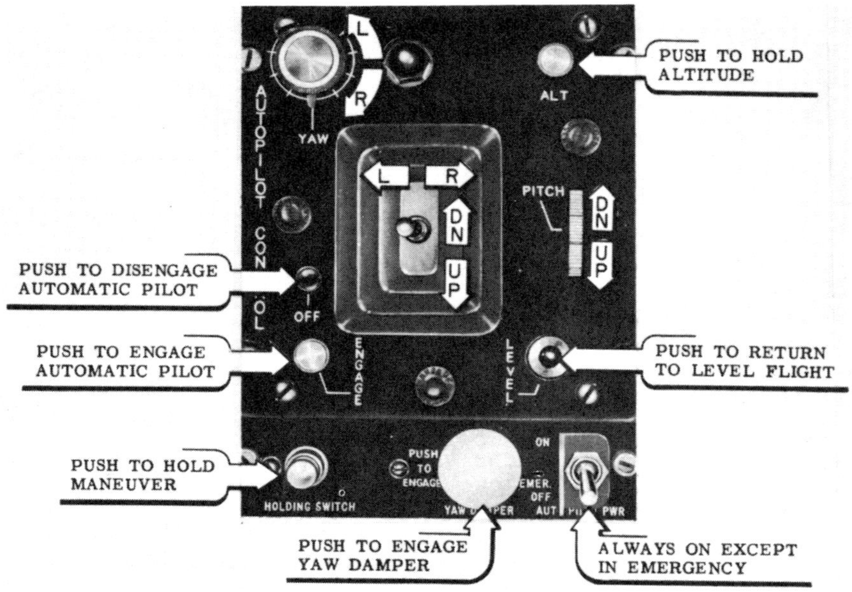

Figure 4-16. Automatic Pilot Controls

switch on the left cabin rail. With the power switch on the MASTER control panel, or the alternate switch on the left cabin rail, set to ON and the mode selector switch on the STA 1 panel set to MASTER or STICK, the READY indicator lights on the STA 1 and STA 2 control panels will glow. When ready for the photographic run, move the operate switch on the MASTER control panel to ON to start automatic continuous camera operation.

CM-1 PHOTO RECORDING UNIT.

This equipment records a counter reading on the center of the strip film of the S-7-S camera at approximately six inch intervals. A synchronized, similar counter and a film speed indicator are mounted in the indicator unit and are photographed simultaneously by the photo recorder camera. The pulse for actuating the photo recorder is initiated by the photo interval timer. The counter readings on the continuous strip film provide reference points which can be correlated with film speed readings, altitude, and other flight instrument information photographed by the photo recorder camera.

AUTOMATIC PILOT—F9F-6P AIRPLANES.

WARNING

Only the yaw damper portion of the automatic pilot is operative.

DESCRIPTION.

The G-3 Automatic Pilot is an electrical and electronic device which automatically maintains the airplane in level flight or at any desired attitude through a range of 65 degrees in climb or dive, 65 degrees in roll, and 360 degrees in azimuth. When the airplane is displaced from its course with the auto pilot engaged, it is returned promptly and smoothly to its original attitude and course. A constant barometric altitude can be maintained by pushing a switch. The auto pilot can be used for maneuvering; it will automatically provide enough rudder for a ball-centered turn, within 1/4 ball, and sufficient up-elevator to prevent appreciable loss of altitude during turns. The auto pilot also incorporates automatic yaw damping. This portion of the system functions to increase the directional stability of the airplane when flying with manual controls and/or auto pilot engaged.

The auto pilot uses both a-c and d-c power. For normal operation, the a-c power is supplied by the No. 2 inverter and is protected by the No. 2 inverter circuit breaker and two G-2 compass a-c fuses. The d-c power for normal operation of the entire auto pilot system, with the exception of the G-2 compass portion, is supplied from the monitor bus and is protected by the auto pilot circuit breaker. The d-c power for the G-2 compass portion is supplied from the essential bus and is protected by the flight instruments circuit breaker and the G-2 compass d-c fuses. The circuit breakers and fuses are located on their respective panels in the right console. An emergency off switch is located on the pilot's control stick grip.

Note

The auto pilot will become inoperative if the No. 2 inverter should fail.

NORMAL CONTROLS.

AUTO PILOT POWER SWITCH.

This switch, located on the yaw damper panel on the right console, controls the output of the control amplifier to the servos, the voltage to all internal relays and the voltage that engages the magnetic clutches for the servos. The switch has two positions, ON and EMER OFF. It should always be left in the ON position except in an emergency, when it is necessary for the pilot to immediately take over manual control of the airplane.

YAW DAMPER SWITCH.

This switch is a large knob located on the yaw damper panel and placarded PUSH TO ENGAGE. When it is pushed in, it permits the auto pilot to operate as a yaw damper only. With the auto pilot power switch set to ON, only the rudder channel is engaged; the flaperon and elevator channels remain synchronized and will provide smooth take-over when the ENGAGE push button switch on the stick-type controller is pressed.

HOLDING SWITCH.

This switch, located on the yaw damper panel, when depressed, holds the airplane in a particular attitude until the LEVEL push button switch on the stick-type controller is pressed to bring the airplane back to a level flight attitude.

STICK-TYPE CONTROLLER.

This unit contains four push button switches, two controls and a miniature stick. The four push button switches, placarded ENGAGE, OFF, LEVEL and ALT., serve to engage the auto pilot, disengage the auto pilot, return the airplane to level flight or hold the airplane to a constant barometric altitude, respectively. The controls marked YAW and PITCH act to make small changes in pitch attitude or heading and the miniature stick is used to maneuver or set up a desired attitude and course. When the auto pilot is engaged, movement of the miniature stick simulates equivalent movement of the control stick and rudder pedals and results in coordinated maneuvers.

Note

The auto pilot should not be considered capable of being overpowered. To take over control in an emergency, set the auto pilot power switch to EMER OFF position.

MASTER DIRECTION INDICATOR.

This instrument supplies a course signal for use by the auto pilot. The caging knob on the MDI is used to establish the initial on course heading on the outer dial.

PREFLIGHT CHECK.

Note

Before turning on the auto pilot, check the manual controls. They should move freely. After this check, center the control surfaces manually.

The auto pilot cannot be engaged with the airplane on the deck while the engine start master switch is ON. To check auto pilot operation on the deck, set the engine start master switch to OFF, ONLY for duration of ground check.

Each of the following checks should be performed as quickly as possible.

a. Check that the No. 2 inverter circuit breaker is pushed in and make sure the auto pilot power switch is set to ON and the auto pilot is disengaged.

b. Wait two minutes for the system to warm up and push the engage push button switch.

Note

The rudder may drift slowly. This is normal and does not indicate any fault in the system. A rapid movement of the rudder may indicate malfunction. When the airplane is on the ground, deflecting the control surfaces cannot change the airplane's attitude and therefore, will not change the gyro output or pendulum, as it would when airborne. Therefore, the rudder and elevators are likely to drift when the airplane is on the ground. Drift does not occur with flaperons.

c. Move the miniature stick to the right, left, forward, and aft and note that the flaperons and longitudinal control surfaces respond in the proper direction.

Note

If the manual stick goes hard over and/or there is no response to course trim, set the inverter changeover switch to #1 INV. Return the switch to #2 INV. & AUTO PILOT and wait two minutes for warm-up. Recheck that gyro is erect.

d. Roll the pitch wheel forward; the elevators should go down as for a dive. Roll the pitch wheel aft; the elevators should come up as for a climb.

e. Turn the yaw knob clockwise; the flaperons should move for a right bank. Turn the yaw knob to center position and then counterclockwise; the flaperons should move for a left bank.

f. Check turn compensation by momentarily displacing the miniature stick in pitch and snapping it to right and left; elevators should go up into climb position. Drift can be expected.

Note

Check the operation of manual controls. The effects of the auto pilot on the control surfaces should be felt definitely, and should require considerable force to overpower them. If the auto pilot is easily overpowered, it is not functioning properly.

g. Set auto pilot switch to the EMER OFF positon.

OPERATION.

BEFORE ENGAGING AUTOMATIC PILOT.

a. Manually trim the airplane.

b. Set the outside dial of the master direction indicator to approximate the heading of the magnetic compass. This is accomplished by pushing in and turning the caging knob on the master direction indicator.

ENGAGING THE AUTOMATIC PILOT.

CAUTION

As the vertical gyro must be erected when the airplane is level, the auto pilot must be readied (steps a., b., c. and d.) when the airplane is in ground status.

When appreciable changes of speed are contemplated while flying on auto pilot, it is recommended that system be disengaged, the airplane be retrimmed and auto pilot be reengaged. This will avoid a sudden and unexpected maneuver when auto pilot is later disengaged. Before engaging the auto pilot, be sure to trim the airplane for attitude and speed at which the airplane is cruising.

Disengage the auto pilot if maneuvers exceeding 65 degrees in pitch or bank are desired.

a. Make certain the No. 2 inverter circuit breaker is pushed in and auto pilot power switch is set to ON.

b. Allow two minutes for the auto pilot to warm up—for the amplifier circuits to be energized and the gyros to be brought up to speed and erected. The auto pilot is now ready to take control of the airplane whenever desired.

c. Push the engage push button switch; if it does not stay down, wait a few seconds and try again.

WARNING

If this check is made before take-off, be sure to push the off push button switch after step c.

Note

The auto pilot cannot be engaged if it is not synchronized or if the power supply is too low. Likewise, if the power supply should fall too low for satisfactory operation, the system will automatically disengage.

d. Move the yaw knob and the pitch wheel to their center positions. Disengage auto pilot by pushing off push button on stick-type controller.

CAUTION

If the No. 2 inverter drops out FOR TWO MINUTES OR MORE due to flame-out or cut-back of the engine to less than 46% rpm (only with low battery), proceed as follows:

ON THE GROUND—when the No. 2 inverter cuts out, the No. 1 inverter is automatically cut in. Set the inverter changeover switch to #1 INV. position and wait seven minutes for the gyro to come to rest. Then set the inverter changeover switch to #2 INV. & AUTO PILOT position and proceed as outlined in steps a., b., c. and d.

IN THE AIR—when the No. 2 inverter cuts out, the No. 1 inverter is automatically cut in. Set the inverter changeover switch to the #1 INV. position. SET THE AUTO PILOT POWER SWITCH TO OFF POSITION. THE AUTO PILOT CANNOT BE USED DURING THE REMAINDER OF THE FLIGHT.

e. Push the engage push button switch when ready to use the auto pilot. The airplane will hold the same attitude and rate of turn it had before engagement. The auto pilot can be engaged with the airplane in any attitude up to 65 degrees in pitch and bank. After rapid maneuvers under manual control WAIT AT LEAST 12 SECONDS BEFORE ENGAGING THE AUTO PILOT. AFTER LOOPING WAIT AT LEAST 60 SECONDS BEFORE ENGAGING.

f. To level the airplane in pitch and bank, momentarily depress the level push button switch. The same result can be obtained by momentarily moving the miniature stick in two axes. With either action, the airplane will assume level flight attitude smoothly, provided it was trimmed properly and attitude and speed have not been changed greatly from the condition to which the airplane was originally trimmed.

g. To level the airplane in pitch only, while retaining bank, momentarily displace the miniature stick in a fore or aft direction.

h. To level the airplane in bank only, while retaining pitch momentarily displace the miniature stick to left or right.

i. To make small changes in pitch attitude or heading, turn the pitch wheel or yaw knob.

CAUTION

Do not swing the yaw knob more than one-half full movement at any one time or the airplane will go into a sharp coordinated turn.

j. To fly a constant barometric altitude, push the ALT push button switch.

CAUTION

This control should be engaged only when the airplane is in level flight. If previously engaged, the altitude control will cut out during dive or climb and reengage automatically when miniature stick is returned to neutral.

k. To dive or climb, push forward or pull aft the miniature stick.

l. To turn, move the miniature stick to left or right. Turns are coordinated with 1/4 ball for all airspeeds. Up-elevator during turns is automatically provided to compensate for the tendency to lose altitude.

CAUTION

Perform maneuvers beyond 65 degrees manually—DISENGAGE AUTO PILOT.

m. When it is desired to hold the airplane's attitude during a maneuver, press and hold down the holding switch, release the miniature stick and when it returns to neutral, release the holding switch. The airplane will continue to fly in the desired attitude. Pushing the level push button switch will return the airplane to level flight.

n. To disengage the auto pilot, leaving the yaw damper in operation, push the OFF push button switch.

o. To disengage the yaw damper, set the yaw damper switch to the OFF position.

Note

When landing, the yaw damper should be disengaged so that the rudder can be easily controlled. The yaw damper or auto pilot will not operate with the airplane's weight on the wheels.

p. To turn off the auto pilot, set the auto pilot power switch to the OFF position.

MISCELLANEOUS.

Chartboard stowage is provided on the right side of the cabin, outboard of the fuse panel. To stow the board, rotate the latch below the cabin rail, set the board in place in the channel just above the fuse panel and rotate the latch to secure the board in position. To remove the board for use, rotate the latch and lift out the board. The map case is located at the aft inboard end of the right console. A relief tube is provided and the personal gear composite disconnect, with connections for oxygen, "G" suit, radio and pilot's suit heat, is installed on the left console.

NAVY

SEE CONFIDENTIAL SUPPLEMENT

CO 01-85FGD-1A

SEE CONFIDENTIAL SUPPLEMENT
CO 01-85FGD-1A

SECTION VII

TABLE OF CONTENTS

SECTION VII SYSTEMS OPERATION

ENGINE OPERATION—GENERAL FLYING.

Tailpipe temperature control is of the utmost importance to the life of the engine; therefore, it is essential that the limiting temperature for any given condition not be exceeded. When operating with the primary (normal) fuel system, the following characteristics are present:

a. After making an increase in power setting, the full tailpipe temperature increase may lag. After the initial increase is noted, the temperature may continue to increase slowly for as long as five minutes. The stabilized temperature should be within limits established for the various power conditions. (See figure 5-3.)

CAUTION

> Normal acceleration temperature must never exceed 800°C (1470°F).
>
> Avoid prolonged operation between 68 and 76% rpm except in emergencies.

b. Rapid deceleration may be made without flame-outs.

c. The engine fuel control unit provides compensation to maintain, within close limits, any selected engine speed throughout broad ranges of altitude and airspeed.

COMPRESSOR BLADE VIBRATION.

A compressor inducer blade resonant vibration peak exists in the 68-76% rpm range. This range should be avoided during all normal operation but may be used for short periods, if required, for carrier landing approaches, instrument approaches or under other short time operating conditions. This range may also be used for any emergency flight condition. (This reference to "emergency" condition should not be confused with operation with the engine fuel system selector switch in the EMER. position during a failure or suspected failure of the primary fuel system.) The frequency of this resonance is such that, although it cannot be felt by the pilot, it will reduce the life of the inducer blades.

ENGINE MALFUNCTIONING.

If, during take-off, flight or landing, malfunctioning of the primary fuel control system should occur, as evidenced by a dropping off of power, decrease of rpm or amber emergency fuel system warning light indicating low fuel pressure in the primary fuel system, the engine fuel system selector switch should be set aft to transfer control to the emergency system. The rpm should then increase to a safe flight value, limited by the topping governors in the fuel pumps. If rpm exceeds that required for take-off, it should be reduced immediately to the take-off value by use of the trottle.

Note

> 100% engine rpm may not be possible on a cold day when the switch has been set to the emergency system position.

After the transfer, throttle sensitivity will be similar to that when operating on the primary system at sea level, except there will be no automatic acceleration or deceleration control. *Accelerations must be made carefully* to avoid excessive tailpipe temperatures due to overfueling. As altitude increases, engine speed for a given throttle movement will increase markedly. If conditions should permit returning operation to the primary fuel control, it may be accomplished by setting the engine fuel system selector switch momentarily to RESET PRIMARY and then to PRIMARY (J48-P-6A) or START & PRIMARY (J48-P-8).

CAUTION

When operating on the emergency fuel system, check tailpipe temperatures carefully.

The emergency fuel system may be used at any time during flight by setting the engine fuel system selector switch to START & EMER. Except under take-off conditions, *the throttle must be retarded to IDLE prior to switchover* to prevent overfueling and possible flame-out. It is also desirable to make transfers from the emergency to the primary fuel system with the throttle at IDLE.

CAUTION

Operation on the emergency fuel system places more severe requirements on the main pumps; the emergency fuel system should not be used for extended periods of operations except *under actual emergency conditions.*

ENGINE ACCELERATING CHARACTERISTICS.

At sea level, the times required for engine acceleration are as follows:

Idle to 100% rpm	8 seconds
65 to 100% rpm	5 seconds
85 to 100% rpm	3 seconds

Certain engine acceleration characteristics which may be experienced during high altitude operation are discussed below.

a. In performance of snap or maximum rate accelerations at high altitude, around 40000 feet, the ideal control setting produces a smooth acceleration throughout the entire engine speed range. However, with the average control setting, some engine roughness is to be expected during the acceleration condition at altitudes above 30000 feet but, up to a point, should not be the cause of undue concern.

b. As control settings become richer, rough operation ranges become apparent. As richness is increased further, the roughness becomes increasingly severe and extends over slightly wider engine speed ranges until the point of rich flame-out is reached. This roughness would be expected to be most severe in the 90-95% rpm range, although another roughness range is sometimes noted at approximately 85% rpm.

c. A setting that causes roughness which comes in and then fades out through the above stated speed ranges, without changing noticeably in frequency and rate of engine speed increase, is acceptable. With this condition, the throttle can be handled in any manner over any engine speed range without concern for flame-out. This roughness can be described as a buzz of fairly high frequency and is more heard than felt.

d. Warning of an approaching flame-out from an overrich control manifests itself in the form of a decrease in frequency and an increase in severity of roughness during the acceleration, together with a slowing, sometimes to zero, of the rate of engine speed rise. The tailpipe temperature is not necessarily affected; it may hold fairly steady or, more likely, tend to drop. As the flame-out is approached, the roughness can be described as changing from a buzz to a heavy stutter, together with a slowing or stopping of the acceleration rate. This condition occurs very close to the flame-out point and, if reached, the throttle should be *retarded immediately* to some intermediate point to prevent flame-out.

e. It has been found in accelerating from intermediate rpm settings, rather than from idle, that roughness will sometimes be more severe than through the same engine speed during snap acceleration from idle. Identification of proximity to flame-out and the remedial action for this condition is the same as above.

f. A decrease in the rate of advance of the throttle will decrease the tendency for flame-out from an overrich control setting and will decrease the severity of an acceleration roughness. Decreasing airplane altitude has the same effect.

IN-FLIGHT REFUELING.

Certain airplanes are equipped for in-flight refueling of both fuselage tanks. This is accomplished by means of a system consisting of a probe, piping and valves and additional venting of the main fuel cells. Fuel flow is regulated principally by control of tanker pump pressures and normally amounts to 1200 pounds per minute on the tanker high pressure pump and 300 pounds per minute on the low pressure or emergency pump. Fuel cell back pressure is relieved by opening the vent valve electrically by means of the in-flight refueling switch on the left cabin rail. Power to operate the vent valve comes from the essential bus. The switch has two positions, OFF and FUEL.

Present tanker aircraft employ a kit which is suspended in the bomb bay on bomb shackles. This kit consists of a fuel cell, a high pressure pump, a low pressure or emergency pump, a reel with 70 feet of hose, on the end of which a drogue is attached and a pylon which may be lowered and raised from the bomb bay to permit the hose to clear the tanker aircraft. The pylon has an opening through which the hose is streamed after leaving the reel, and two lights, an amber light to indicate that the tanker is ready to deliver fuel and a green light to indicate that fuel is flowing.

The kit is operated by the following procedures: The pylon is lowered and the hose reeled out to its full 70 foot length. In this condition, the tanker is ready to deliver fuel and the amber light should glow. Automatic controls prevent the flow of fuel while the hose is extended beyond a 40 foot length. This automatic control also starts fuel flow when the hose has been forced back to a 40 foot or less length by the fighter. Initial fuel flow is generally pumped by the high pressure pump at a 1200 pounds per minute rate. This fuel flow is maintained until the tanker shifts to emergency pump for topping off. The emergency pump delivers 300 pounds per minute and will stall out when the tanks are full. The green light on the pylon will glow when fuel is flowing.

It should be noted that tanker automatic controls are designed to prevent a high pressure blow-off (separation of probe from drogue due to fuel pressure, accompanied by considerable loss of fuel) or damage to the receiver airplane's fuel cells should the cells be filled by the high pressure pump alone. However, it is considered advisable to notify the tanker by radio when the fuselage tanks are within 400 pounds of being full and to top off on the emergency pump.

GROUND CHECKS.

With the battery switch on, actuate the in-flight refueling switch in the cabin and have someone listen in the plenum chamber blow-in doors to ensure that the solenoid which opens the vent valve is operating.

Conduct a ground fuel integrity check. This check is best accomplished by attaching the tanker drogue to the probe and actually fueling the main cells from the tanker aircraft, using the same procedures as recommended when airborne. If this is not possible, a tanker drogue may be adapted to the hose of a fuel truck and the main cells filled through the probe. It is advisable to remove the nose shell prior to a fuel integrity check to facilitate the detection of fuel leaks in the piping through the nose section. The piping through the nose wheel well can be readily inspected from that area. If there is no fuel leakage in these areas, the aircraft is ready for aerial refueling.

FLIGHT PROCEDURES.

Note

> Only the fuselage tanks may be refueled. Therefore, if a refueling contact is planned, the flight prior to refueling should be made using only the fuselage fuel, reserving the wing fuel for later use. If fuselage fuel gets low, it should be maintained at approximately 300 pounds by intermittently transferring fuel from the wing tanks.

Rendezvous with the tanker, using the AN/ARA-25 homing equipment or the AN/ARN-6 radio compass, tuned to the tanker's homing equipment. After rendezvous has been effected, close in on the tanker from astern until flying in steady state with the probe approximately 10 feet aft and below the tanker drogue. This will place the airplane clear of the tanker's prop wash. The highest speed the tanker can maintain with the drogue extended will allow the most comfortable attitude for the receiver airplane. With present configurations, this speed is approximately 200 knots IAS. This is approximately the speed of the F9F-6 at which the rpm required is at a minimum. If lower speeds are desired, power must be added. Observe the lights on the tanker, an amber light indicating that the tanker is ready to deliver fuel. Set the in-flight refueling switch to FUEL. Shift to 100% oxygen. Add 4-5% rpm and allow the fighter to accelerate slowly but steadily, flying the probe into the funnel surrounding the drogue.

Should you overrun the funnel without engaging it, reduce power, resume starting position and make another approach. Airflow around the fuselage will prevent a collision with the drogue. After the probe enters the funnel, it will slide into the drogue and seat itself. After the probe has been seated in the drogue, adjust the power setting to close in on the tanker slowly, allowing the hose to reel in while preventing it from developing a loop. When the hose has been reeled in from its original 70 foot length to 40 feet, the green light will glow on the tanker pylon to indicate that fuel is flowing. Fly formation on the tanker, bearing in mind that a considerable power setting change may be required, depending on weight of fuel delivered, as the tanker becomes lighter and the receiver heavier. Fuel will continue to flow until the fuselage tanks are full (as long as the hose is maintained at a 40 foot length or less). Watch the fuel quantity gage and when it shows the tanks to be within 400 pounds of full notify the tanker by radio to shift to emergency pump. The emergency pump will stall out and fuel will stop when the tanks are full. In this condition, the receiver can ride as a parasite, using tanker fuel while maintaining full fuselage tanks.

To detach from the tanker reduce rpm 4-5% and allow a steady rate of opening from the tanker to develop. This rate must be carefully controlled to prevent both too slow a withdrawal, which makes separation difficult, and too rapid a withdrawal which will damage the tanker reel mechanism. A 5-7 knot differential is considered ideal. At this stage, rise up under the tanker until the drogue is in approximately the same relative position to the tanker as it was prior to contact. On a perfect separation, the probe will break clean without contacting the funnel. After separation, turn the in-flight refueling switch to OFF and return to normal oxygen.

WARNING

> The in-flight refueling switch should be set to OFF immediately. Failure to do so will result

in the loss of fuel tank pressure which may result in reduced fuel flow and, in extreme cases, a flame-out.

All electronic and electrical equipment may be operated during refueling. Use of the AN/ARC-27 is particularly recommended, except under conditions of radio silence, since the receiver cannot be observed by the tanker during refueling.

SECTION VIII
CREW DUTIES

NOT APPLICABLE

SECTION IX

TABLE OF CONTENTS

INTRODUCTION.

This section contains only those procedures which differ from, or are in addition to, the normal operating instructions covered in Section II. Any discussion relative to systems operation is covered in Section VII. Where repetition occurs, it is mainly for emphasis, clarity, or continuity of thought.

INSTRUMENT FLIGHT PROCEDURES.

GENERAL.

This paragraph does not attempt to cover the general technique of flying weather, but is presented to acquaint the pilot with the qualities, capabilities, and limitations of the airplane under instrument flight conditions.

A successfully conducted instrument flight, especially in jet aircraft, requires thorough preflight planning, current instrument proficiency on the part of the pilot, and adequate instrumentation for blind take-off, orientation, approach, and landing.

Though designed primarily as a day fighter and interceptor, this aircraft is capable of comfortable instrument flight, provided the restrictions prescribed below are observed. The instrument flight handling characteristics and stability about the three axes are adequate; instrumentation is good. Defogging and defrosting are provided for the windshield. There are no provisions for exterior de-icing.

The vhf (or uhf) receiver-transmitter provides long range, static-free communication. The low frequency and uhf radio compasses and the homing receiver are used in navigation, orientation and let-downs. The radio altimeter and automatic pilot provided in F9F-6P airplanes are further aids in instrument flight.

A reduction in range should be anticipated in weather flying for the following reasons:

a. Dog-legging to avoid thunderstorm areas or to navigate on airways.

b. Holding in the stack, awaiting clearance for approach.

c. A decreased rate of descent may be required to conform with approach procedures or local conditions.

d. An unsuccessful range orientation or missed approach.

INSTRUMENT TAKE-OFF.

PRIOR TO TAKE-OFF.

Because of the extremely high fuel consumption of jet engines; the pre-take-off check list must be completed as far as possible prior to starting the engine. This can be accomplished in full, with the exception of the control systems check and engine check.

a. Connect an external source of electric power.

b. Energize all normal battery circuits and check all trim functions, gyro instrumentation, and pitot heater.

c. Complete all necessary radio checks with the tower and traffic control, receiving latest altimeter setting. Advise tower of intention to stand by in run-up position until final clearance for departure is received, at which time engine will be started, airplane taxied, and take-off accomplished in a very short time.

d. When ARTC clearance is received, start engine, check engine and control systems, set flying tail off, yaw damper on, and taxi immediately to the runway.

e. Tune in all radio aids neded just prior to take-off, avoiding overloading of battery.

f. Check pitot heater on, cabin heat and defrost on and hot, and make final check of cabin just before take-off.

g. Line up on runway, set gyro compass to runway heading, and set the attitude indicator of the gyro horizon in desired position.

TAKE-OFF.

a. Hold brakes and advance throttle, making final check of engine instrumentation.

CAUTION

Do not advance throttle to full rpm while holding brakes when ice, snow or rain are present on the runway. The large amount of thrust may cause the tires to slide uncontrollably over the runway surface.

b. Release brakes and maintain gyro compass heading by use of brakes and/or flaperon until rudder control becomes effective.

c. As soon as airborne, maintain take-off attitude by gyro horizon to 500 feet, making full panel cross reference of flight instruments.

d. Wheels up as soon as practicable.

e. Flaps up at 1000 feet, and accelerate to best climbing airspeed.

WARNING

In raising and lowering the flaps and slats, a slight change in longitudinal trim occurs; this should be anticipated in order to maintain the established climb attitude.

INSTRUMENT CLIMB.

a. Climbing turns should be standard rate. Do not exceed banking angles of 30 degrees.

b. Avoid flight in icing levels, since no exterior de-icing equipment is provided. Use pitot heat and windshield defrosting as required.

INSTRUMENT CRUISING FLIGHT.

a. Trim aircraft for cruising flight, using gyro compass and gyro horizon for quickest and easiest reference. Fly full panel instruments for a true picture of the aircraft's attitude and performance, glancing around the cabin occasionally to avoid overconcentration on the flight instruments. This will help to reduce fatigue and eyestrain over long periods of time.

b. Yaw damper on HIGH GAIN to eliminate lateral-directional oscillations; in rough, turbulent air, particularly at low flight speeds, the motion of the rudder pedals may be annoyingly large, in which case the LOW GAIN position will make it more tolerable.

c. The flying tail is particularly effective in high speed cruising flight; the high effectiveness and light control forces required result in a high degree of sensitivity. Avoid abrupt stick motions to prevent overcontrolling.

d. The speed range in smooth air will depend on altitude, range, and pilot preference. In turbulent air, speeds above 350 knots IAS present undesirable control characteristics and pilot discomfort, as well as imposing excessive "g" forces on the airplane. (See Flight in Turbulence and Thunderstorms paragraph.)

e. Navigation by use of the AN/ARN-6 radio compass may prove unreliable at high altitudes, in or near storms, or along mountain ranges and shore lines. Low frequency navigation is also highly susceptible to electrical and precipitation static. Therefore, do not rely solely on the low frequency ADF COMP. position, but cross check against radio range signals and AN/ARA-25 bearings. If reception on the ANT. or COMP. position is unsatisfactory, the LOOP position may be selected to give clearer reception.

The LOOP position may also be used for direction finding more reliably than the COMP. position during unfavorable conditions by rotating the DF needle until an aural null of approximately 10 degrees spread is received. The spread may be varied by varying the volume control. Be sure to solve the 180 degree ambiguity when using this method.

DESCENT.

The rate and speed of descent will depend on several factors such as terminal weather, fuel remaining, turbulence, icing, desired altitude over destination, radio facility, etc. Slow airspeed descents will decrease effective range, as well as lower the efficiency of the windshield and canopy defrosters when using lower engine rpm. DEFROSTING SHOULD BE TURNED UP FULL, WELL BEFORE STARTING DESCENT. If increased effectiveness of the defroster is desired during a slow airspeed descent, increase rpm and use speed brakes to hold down airspeed.

a. In smooth air, the Optimum Descent charts of Appendix I may be used for best range. In a high speed descent, restrict bank angle to 30 degrees to prevent entry into a winding spiral. In a high speed, optimum range descent, do not lower flaps or speed brakes, as they reduce the aerodynamic efficiency of the airplane.

b. In turbulence, storms, or mountainous terrain, a slower let-down airspeed is preferred because:

The "g" forces imposed on the airframe by gust loads are lessened;

The airplane can stay close to the range station or radio fix during descent;

Better ability to avoid storm centers and resultant regions of severe turbulence.

Note

A high rate of descent may easily be established at slow airspeeds by extending speed brakes and reducing power. An IAS of 250 knots is recommended for high performance let-downs.

HOLDING.

Use approved holding procedures. If considerable delay is anticipated in the stack, it is advisable to request traffic control clearance to maintain a high altitude for fuel conservation. The Maximum Endurance charts and Fouled Deck Endurance table of Appendix I will be helpful.

INSTRUMENT APPROACHES.

Following the descent and arrival over the high cone of the radio range or fix, the pilot may elect to complete a radio range approach, or he may came in by GCA (Ground Controlled Approach), depending on facilities available, local conditions, and pilot preference.

The yaw damper should be on and the flying tail off (manual elevator control). On F9F-6P airplanes, the auto pilot should be off, and the radio altimeter on, if desired.

WARNING

On airplanes not equipped with the Lear yaw damper, lateral-directional control will not be as effective as with the damper installed, especially in turbulent air.

a. Radio range let-down and approach. If no specific let-down and approach procedure for jet aircraft exists at the station, use the standard range approach charts. Note well the local restrictions on terrain, since your comparatively higher let-down speed may carry you out of the safe areas. On passing the high cone, establish the required rate of descent on the outbound heading. Extend speed brakes and reduce airspeed to 175 knots. Level off for procedure turn, lowering flaps and gear (speed brakes will retract automatically), maintaining airspeed. Complete descent inbound as required, level off, and trim to fly at 150 knots. On passing the low cone, reduce airspeed to 130 knots and descend to minimum altitude.

If visual contact is not established in required time, perform a missed approach procedure. Advance throttle to 100% rpm and raise landing gear; flaps up at safe airspeed and altitude.

b. Ground controlled approach (GCA). The vhf and uhf sets give exceptionally good static-free communication under most conditions. Precipitation and storms can affect the GCA operator's scope, however, by causing intermittent loss of contact and cluttering the scope with precipitation echoes.

Once contact is established with the GCA controller at the specified position, he will direct the pilot to execute certain turns for positive identification, and from then on will advise the pilot what course and altitude to take, how fast to turn, climb or descend, etc. A smooth basic instrument flying technique goes a long way toward a successful GCA approach.

ICE, SNOW, AND RAIN

Ice and snow will adhere to the leading edges of the wings and tail, and the forward portion of the external stores. The ultimate effect will be a decrease in airspeed, increase in power required, increase in stalling speed, and a reduction in range. Runways covered with ice, snow or rain present a hazard in take-offs and landings due to skidding and loss of traction. Therefore, use rudder and flaperons for directional control as much as possible. Furthermore, radio static interference, ADF wandering, cluttering of the GCA scope, or intermittent loss of GCA contact can result from precipitation.

Icing levels should be avoided, as well as fog and clouds at near freezing temperatures, since no de-icing equipment, other than windshield and canopy defrosting is provided. In flight through rain, visibility is better through the side panels of the windshield; decreasing airspeed will improve visibility and the use of approved water repellents applied to the windshield will also help.

WARNING

Icing will cause the airplane to stall at a much higher airspeed, due to a reduction in aerodynamic efficiency and increase in weight.

FLIGHT IN TURBULENCE AND THUNDERSTORMS.

The airplane should not be operated in conditions of severe turbulence such as are found in thunderstorms and well developed cumulus clouds. However, if flight in severe turbulence cannot be avoided, the yaw damper shall be in operation and the range of airspeeds given in Section V are recommended.

The best method of penetration lies in the early establishment of power setting and pitch attitude on the gyro horizon for level flight at a safe penetration airspeed. By maintaining this constant pitch attitude and power setting, a relatively constant airspeed will result throughout the storm. Chasing the airspeed indicator will result in considerable overcontrolling and extreme attitudes, due to the lag within the instrument, as well as the constantly changing barometric pressure differentials encountered in storms. This barometric fluctuation can also cause the altimeter to give unreliable readings by several thousand feet.

APPROACHING AND ENTERING THE STORM.

Make all possible preparations before entering the storm, in order to have more time for flying the airplane within the storm.

a. Set yaw damper to LOW GAIN.

b. Flying tail off (manual elevator control).

c. Auto pilot off (F9F-6P only).

d. Pitot heater on, defroster on.

e. Reduce power and airspeed.

f. Set gyro instruments.

g. Turn up cabin lights to avoid blinding effect of lightning flashes.

h. Tighten seat belt and shoulder harness.

i. Turn off all unnecessary radio equipment and reduce volume on others.

j. Keep flaps and speed brakes up.

k. Trim airplane to fly power setting and pitch attitude; banking angles should be kept very small. Avoid overcontrolling by riding with the storm a certain amount when correcting attitude.

l. Pre-stall buffeting can become masked by the effects of turbulent flight at high altitudes. Therefore, maintain correct pitch attitude and power setting for safe airspeed.

m. Expect lightning, turbulence, precipitation, icing, and radio interference.

NIGHT FLYING.

The night flying comments presented in this paragraph are given in addition to the normal preparations taken for instrument flight.

PRIOR TO TAKE-OFF.

a. Check interior lighting and adjust to low intensity for dark adaptation. Use glare shield, when provided, for reduction of canopy reflections and glare.

b. Check wing and tail lights, fuselage lights, and formation lights.

c. Have a flashlight available.

DURING FLIGHT.

Adjust cabin and instrument lights to desired intensity.

COLD WEATHER OPERATION.

Icing conditions have already been discussed in the Instrument Flight Procedures paragraph. Cold weather post-flight preparations are not as critical in jet powered airplanes as they are in recipricating engine airplanes, since there is no need for oil dilution.

BEFORE STARTING ENGINE.

a. Blow hot air into the air inlet ducts with a portable heater to prevent damage to the starter mechanism.

b. Apply heater hose to cabin and toward canopy seals.

c. Remove all exterior covers, including pitot cover; remove all ice and snow from fuel and oil tank vents, intake, drainage, and ventilation ducts; and all airplane surfaces and moving parts.

d. Perform a complete visual inspection of the exterior of airplane; check for system leakages, tire underinflation, etc.

WARNING

Dangerous stalling speeds and aerodynamic characteristics result when any snow, ice, or frost are present on the wings and fuselage. Do not attempt a take-off with even the lightest coating of frost on the wings.

e. Check battery to see that electrolyte is not frozen.

f. Have proper external power source available prior to, and during starting of engine.

g. Check for use of proper grade cold weather fuel and lubricants.

AFTER STARTING ENGINE.

Note

When operating in an ambient air temperature of −10°C (+15°F) or lower at sea level, the tailpipe temperature indication may exceed the tailpipe temperature limitation during take-off at 100% rpm. This higher temperature reading is caused by a tendency of the engine during low ambient temperature conditions to run at a somewhat higher tailpipe temperature and also by indicator inaccuracy due to exposure to the cold. If during a take-off at 100% rpm on a cold day the tailpipe temperature appears to exceed this limit (after allowing for possible indicator error), engine rpm should be reduced so that the tailpipe temperature falls within the limitation. The resultant thrust of this reduced rpm will still be greater than or equal to that obtainable under standard day, 15°C (59°F), conditions when at 100% rpm.

a. Check for proper oil pressure.

b. Check all flight controls for full travel and unrestricted movement.

c. Run through a complete cycle with flaps, trim controls and speed brakes several times to assure proper operation.

d. Use only the minimum electrical equipment needed when generator is not charging; battery performance and life are greatly reduced during cold weather operations.

e. Inspect all instruments for proper operation.

f. Operate canopy to assure freedom of movement and locking.

g. Cabin heat and defrost full on; pitot heat on if required.

h. Exercise caution while running up on snow or ice, as chocks will frequently slide.

TAXIING AND TAKE-OFF.

a. Avoid taxiing on ice or snow; tow airplane to take-off position, when possible. If airplane is taxied, allow for a greater braking distance.

b. Avoid icing of airplane surfaces due to melted snow or ice from the jet blast of preceding airplanes.

c. A full power engine check should not be made on snow or ice because of possible slippage; make final engine check during first part of take-off roll.

d. Use brakes carefully, if required during take-off run, to avoid skidding on ice or snow covered runways; use rudder and flaperons for directional control.

e. After take-off from a wet snow or slush-covered airfield, operate the landing gear, flaps, and speed brakes through several complete cycles, at a safe airspeed and altitude, to prevent them from freezing in the retracted position. Gear and flap operation may be slower than normal in cold weather, due to increased viscosity of hydraulic fluid and lubricants.

f. After take-off, set pitot heat switch to ON.

LANDING.

a. Pump brakes several times during approach to free them of ice accumulation or binding.

b. Allow for a change in glide angle due to denser air, increased thrust, and lower TAS.

c. Turn off all unnecessary electrical equipment and keep generator charging whenever possible.

d. Hold nose wheel off runway as long as possible on ice, slush, or snow, to decrease airspeed without using brakes. After nose wheel is on the ground, use brakes sparingly and intermittently. Use flaperons and rudder for directional control. Braking should not be attempted on scattered patches of ice or snow.

e. Pitot heat off.

BEFORE LEAVING AIRPLANE.

a. Leave canopy slightly open to prevent cracking or frosting.

b. Moor airplane firmly and apply wheel chocks.

c. Install all protective covers and plugs required, including pitot cover.

d. Remove battery, if temperature warrants it.

e. Fill fuel tanks to minimize condensation; drain fuel sumps regularly.

HOT WEATHER AND DESERT OPERATION.

All exposed metal surfaces can cause burns after exposure to hot sun. Use caution, and wear protective clothing when required.

BEFORE STARTING ENGINE.

a. Perform a complete visual inspection of the exterior of airplane, checking for system leakages, blistering, deterioration, sand contamination, tire overinflation etc.

b. Remove all protective covers and plugs, including pitot cover; check cleanliness of any screens or filters required.

c. Make all possible ground checks prior to starting engine.

d. Adjust air conditioning as desired.

AFTER STARTING ENGINE.

a. Make ground run-up as short as possible.

b. Do not run-up in direction of other airplanes, personnel, or ground installations.

TAXIING AND TAKE-OFF.

a. Use brakes sparingly in taxiing to prevent overheating; use flaperons and rudder for directional control.

b. Do not attempt take-off in a dust or sand storm; avoid take-off in the wake of another airplane on sandy or dusty airfields.

c. Expect slower acceleration and longer take-off distance, due to lower density of air and decreased thrust at all power settings. TAS will be greater when IAS is normal.

d. Check tailpipe temperature frequently to avoid overheating during take-off and climb; adjust power and airspeed accordingly.

e. Adjust air conditioning as desired.

LANDING.

a. Descent into areas of high temperature and humidity may cause very abrupt foggings of the canopy and windshield; this may also occur at the time the canopy is opened at low altitudes. Therefore, turn defrosters on full, well before starting descent.

b. Expect a change in glide angle, a faster approach, and a longer ground roll, due to lower density of air, decreased thrust and higher TAS.

c. Use brakes sparingly after touch-down to prevent over-heating and subsequent failure; use rudder and flaperons for directional control.

BEFORE LEAVING AIRPLANE.

a. Park airplane cross wind, if possible, to minimize effects of blowing sand and dust.

b. Leave canopy slightly open to permit air circulation if no excessive sand or dust is blowing.

c. Install all protective covers, plugs, and screens required.

d. Fill fuel tanks to minimize condensation of moisture; drain sumps regularly.

NAVY

APPENDIX I
OPERATING DATA

TAKE-OFF DISTANCE CURVES.

Take-off distances are presented graphically in figures A-5 and A-6 for the J48-P-6A and J48-P-8 engines respectively. Both ground run and total distance from brake release to several heights are given, based on the normal take-off technique described in the Take-off paragraph of Section II. These curves are good for configurations A through F. The fuel allowance for start, taxi, take-off and acceleration to climb speed is 480 pounds for the J48-P-6A engine and 520 pounds for the J48-P-8 engine. Following are 2 examples for the airplane with the J48-P-6A engine.

EXAMPLE NO. 1.

Assume that an airplane has a gross weight of 18500 pounds. The runway altimeter indicates 2500 feet and the runway air temperature is 65°F. There is a 20 knot head wind. The runway length is 6000 feet and there is a 100 foot obstacle 7500 feet from the point where the ground run is begun. Enter the temperature scale in figure A-1 at 65°F (18°C) and move up vertically to an altitude of 2500 feet and then horizontally over to a weight of 18500 pounds. Then move down to an obstacle height of 100 feet and horizontally left to the head wind correction grid. Follow the nearest guide line to a head wind velocity of 20 knots. Then move horizontally left again and read 6800 feet total distance to clear the obstacle. To obtain the ground run for this take-off after moving to the proper weight of 18500 pounds, continue vertically down, without stopping at an obstacle height, to the vertical head wind grid. Follow the nearest guide line to a head wind of 20 knots and then move straight down and read a normal take-off run of 4330 feet.

EXAMPLE NO. 2.

Assume the runway altimeter shows 5000 feet, the runway air temperature is 80°F, the head wind is 0, the runway length is 10000 feet, and there is a 100 foot obstacle 11000 feet from the start of the runway. Entering figure A-1 at the temperature and altitude, it is seen that the recommended limit take-off weight is 17900 pounds. At this weight, the ground run will be 7450 feet and the distance to clear 100 feet will be 12600 feet. The actual obstacle conditions of 100 feet at 11000 feet from start of runway, limit the take-off weight to 17000 pounds.

FOR OTHER OPERATING DATA

SEE CONFIDENTIAL SUPPLEMENT

CO 01-85FGD-1A

TAKE-OFF CURVES

HARD SURFACE RUNWAY

MODEL(S): F9F-6 ENGINE(S): J48-P-6A

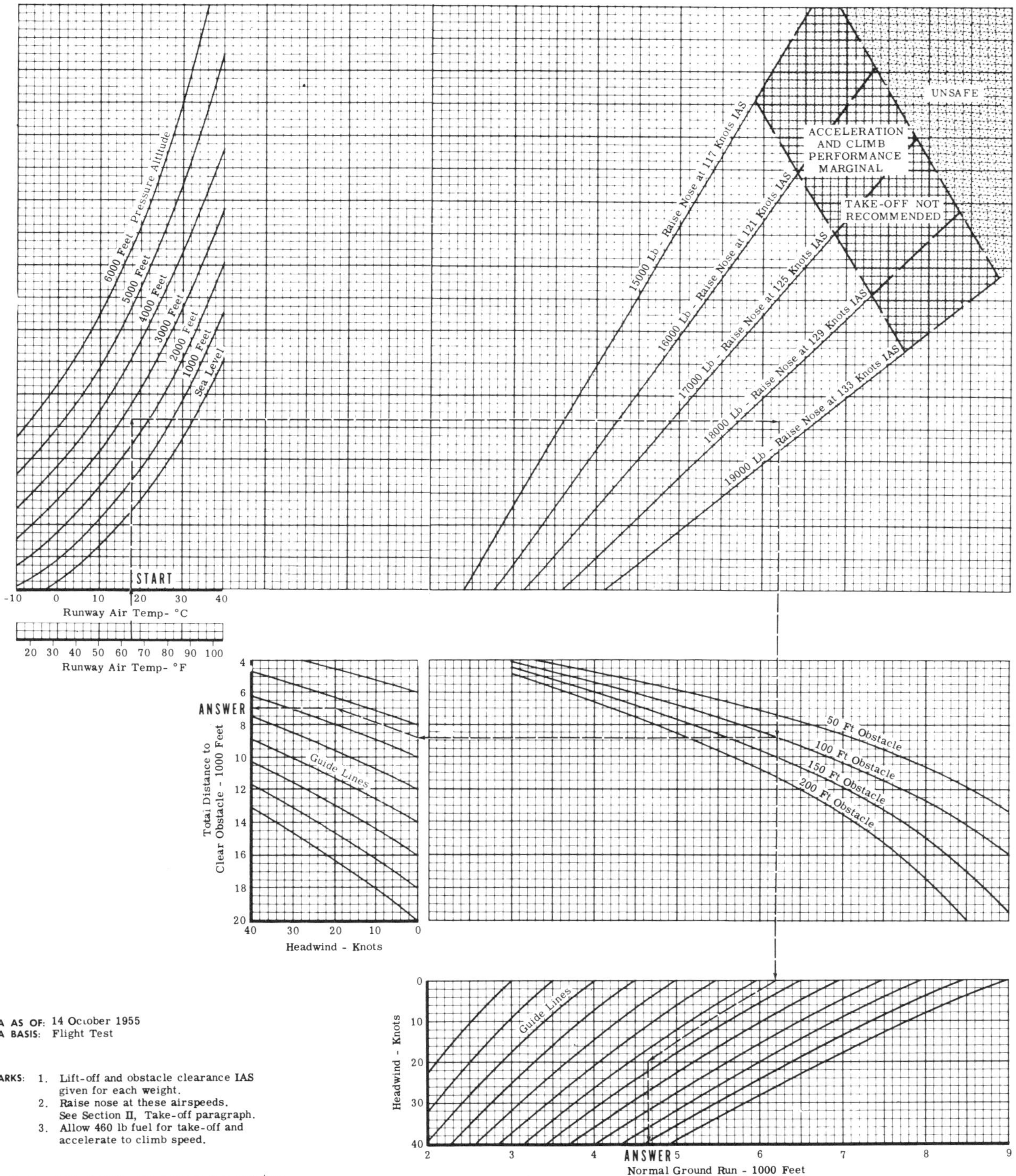

DATA AS OF: 14 October 1955
DATA BASIS: Flight Test

REMARKS:
1. Lift-off and obstacle clearance IAS given for each weight.
2. Raise nose at these airspeeds. See Section II, Take-off paragraph.
3. Allow 460 lb fuel for take-off and accelerate to climb speed.

FUEL GRADE AND DENSITY: MIL-F-5624 (JP-3 or JP-4) 6.5 Lb/Gal
MIL-F-5572 (AVGAS) 6.0 Lb/Gal

Figure A-1. Take-off Distance Curves—J48-P-6A Engine

TAKE-OFF CURVES

HARD SURFACE RUNWAY

MODEL(S): F9F-6, -6P **ENGINE(S):** J48-P-8

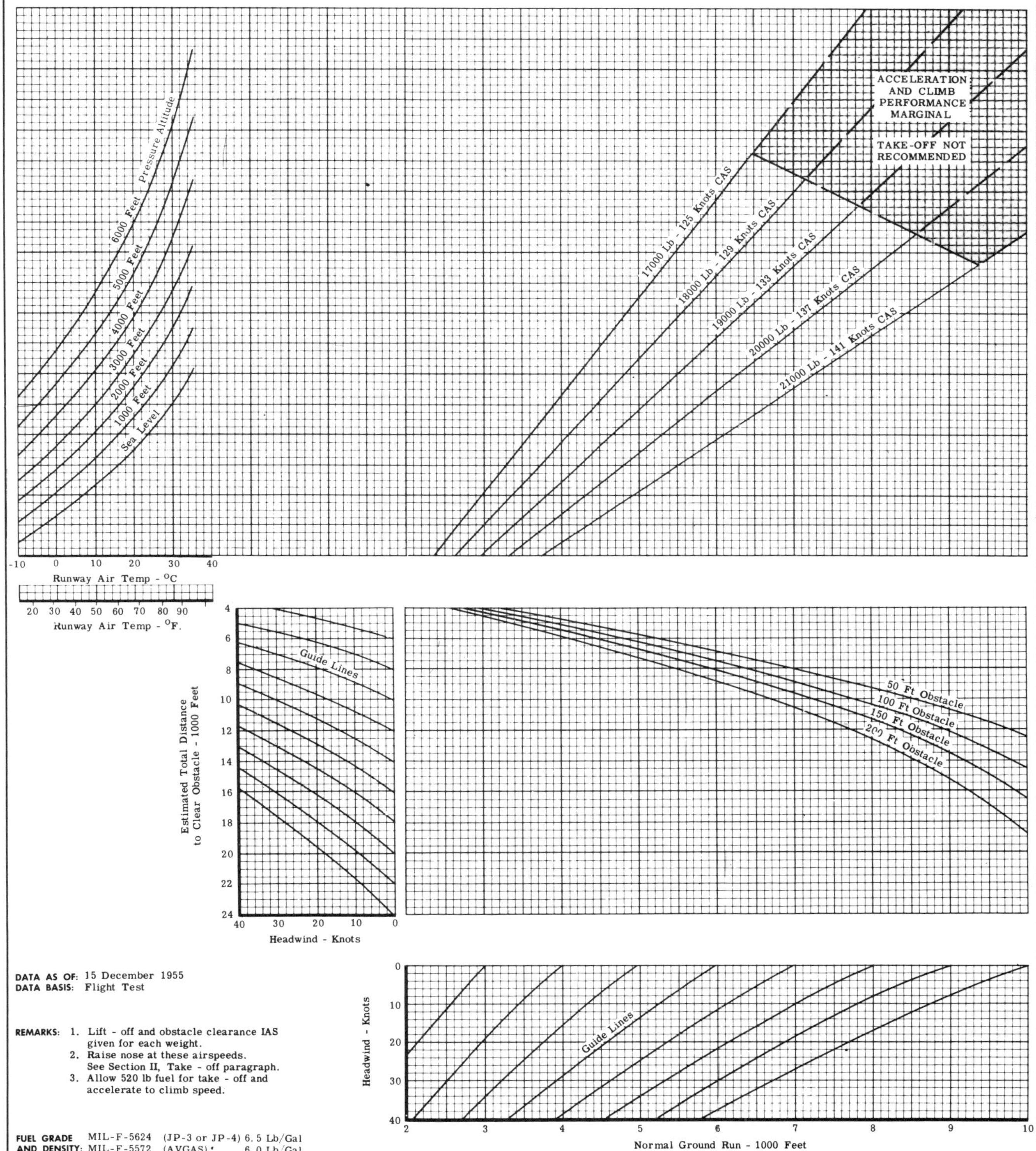

DATA AS OF: 15 December 1955
DATA BASIS: Flight Test

REMARKS:
1. Lift - off and obstacle clearance IAS given for each weight.
2. Raise nose at these airspeeds. See Section II, Take - off paragraph.
3. Allow 520 lb fuel for take - off and accelerate to climb speed.

FUEL GRADE AND DENSITY: MIL-F-5624 (JP-3 or JP-4) 6.5 Lb/Gal
MIL-F-5572 (AVGAS) 6.0 Lb/Gal

Figure A-2. Take-off Distance Curves—J48-P-8 Engine

AN 01-85FGD-1
26 FEBRUARY 1957

FLIGHT HANDBOOK INTERIM REVISION No. 21

Navy Model F9F-6-6P Aircraft

PUBLISHED BY DIRECTION OF THE CHIEF OF THE BUREAU OF AERONAUTICS

Of paramount interest to pilots. To be read by all pilots operating these aircraft

1. Cancellation: None

2. Purpose: This revision sets forth the procedures to be used for the single point ejection system.

3. Instructions: Add the following procedures to the paragraph entitled Emergency Escape in Section III:

EMERGENCY ESCAPE

Emergency escape from this airplane after incorporation of F9F Aircraft Service Change No. 430 should be made by using the ejection seat with the following procedure.

a. REDUCE AIRSPEED IF POSSIBLE.

b. PUSH CANOPY JETTISON CONTROL HANDLE. Pull Handle inboard and push down. (initial pull of the handle will be hard) This will jettison the canopy regardless of its position. The radio and oxygen tube is pulled off the console nipple fitting to which it is attached and the radio lead severed by the force of ejection when the seat has traveled upward a distance of about 12 inches. (the "G" suit hose connection is also separated during the initial upward travel of the seat)

WARNING

Clearance between canopy and pilots head is critical. Make sure head is below normal operating path of canopy before actuating canopy jettison control handle.

WARNING

Ejection through the canopy is recommended only in those cases in which the aircraft is in uncontrolled flight, or the quickest possible escape is necessary; at all other times, the canopy should be jettisoned prior to ejection. Under no circumstances should the pilot attempt to eject with the canopy in the open position. It is recommended that take offs and landings be made with the canopy in the closed position.

c. PLACE FEET ON FOOTRESTS

NOTE

The seat does not lower but remains in a fixed position when the canopy jettison control handle is pushed. The pilot may have to lift his feet up onto the footrest if the seat is in an elevated position.

d. ACTIVATE EMERGENCY OXYGEN SUPPLY. This action is necessary only at altitudes above 20,000 feet.

e. PULL FACE CURTAIN. While continuing to control aircraft as much as possible, reach up with one hand and locate face curtain; then, sitting erect, with head back against headrest and muscles tensed to absorb ejection forces, reach up with the other hand and pull face curtain until fully extended. This will automatically lock the inertia reel and fire the ejection seat catapult.

NOTE

Catapult ground lock handle must be in the flight position (down and aft) to unlock face curtain handle and permit ejection when desired.

f. SEPARATE FROM SEAT. Immediately following ejection, release face curtain, (lap belt will open automatically in 3/4 sec.) and roll forward out of seat if necessary.

NOTE

If time permits, the oxygen tube and radio leads should be disconnected at the alligator clip coupling prior to ejection. When using ejection seat for emergency escape at 2000 feet or below, it is recommended that the oxygen tube and radio leads be disconnected prior to ejection. This procedure will shorten the time required for separation from the seat after ejection.

g. PULL PARACHUTE RIP CORD. If above 3000 feet at time of separation from seat, delay at least five seconds before pulling rip cord. This delay will preclude any possibility of collision between the seat and parachute and willalso result in a lower parachute opening shock. When the escape is accomplished at high altitude, it is recommended that the pilot free fall to an altitude of approximately 18,000 feet before opening his parachute in order to avoid the effects of anoxia and cold.

WARNING

Do not pull rip cord while in seat.

END

AN 01-85FGD-1
17 JANUARY 1957

FLIGHT HANDBOOK INTERIM REVISION No. 20

Navy Model F9F-6-6P Aircraft

PUBLISHED BY DIRECTION OF THE CHIEF OF THE BUREAU OF AERONAUTICS

Of paramount interest to pilots. To be read by all pilots operating these aircraft

1. Cancellation: None

2. Purpose: This revision sets forth procedures to be followed in the event of malfunctioning of the nutcracker switch in the landing gear safety circuit.

3. Instructions: The following caution is to be added to the Flight Handbook AN 01-85FGD-1, Navy Model F9F-6,-6P, dated 15 April 1956. Pg. 58 following the paragraph entitled LANDING GEAR:

CAUTION

> In the event the landing gear handle can not be raised after take off except by manual operation of the landing gear down lock solenoid release, the flight should be aborted.

4. Reasons for the above instructions are as follows:

a. Failure of the landing gear down lock solenoid release may indicate an inoperative nutcracker switch.

b. Malfunctioning of the nutcracker switch in the landing gear safety circuit can prevent satisfactory ignition during air starts.

END

AN 01-85FGD-1
20 SEPTEMBER 1956

FLIGHT HANDBOOK INTERIM REVISION No. 19

Navy Model F9F-6 -6P Aircraft

PUBLISHED BY DIRECTION OF THE CHIEF OF THE BUREAU OF AERONAUTICS

Of paramount interest to pilots. To be read by all pilots operating these aircraft

1. Cancellation: NONE

2. Purpose: This revision sets forth revised take-off characteristics for F9F-6, -6P aircraft with the J48-P-8 engine installed.

3. Delete figure A-2, page 165 of Flight Handbook AN 01-85FGD-1 dated 15 April 1956, and substitute the curve on page 2 of this revision.

TAKE-OFF CURVES

HARD SURFACE RUNWAY

MODELS: F9F-6, -6P **ENGINE: J48-P-8**

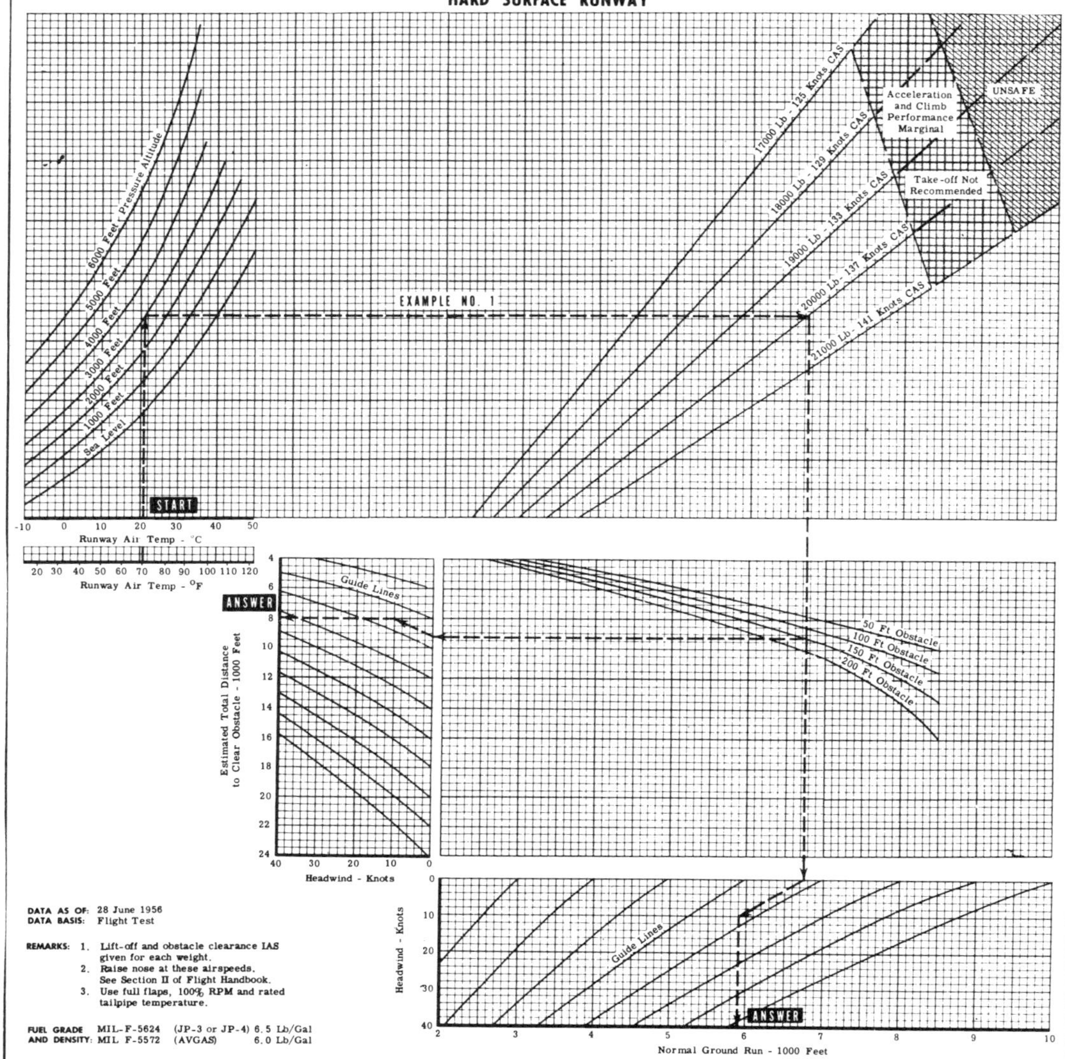

DATA AS OF: 28 June 1956
DATA BASIS: Flight Test

REMARKS:
1. Lift-off and obstacle clearance IAS given for each weight.
2. Raise nose at these airspeeds. See Section II of Flight Handbook.
3. Use full flaps, 100% RPM and rated tailpipe temperature.

FUEL GRADE AND DENSITY: MIL-F-5624 (JP-3 or JP-4) 6.5 Lb/Gal
MIL F-5572 (AVGAS) 6.0 Lb/Gal

EXAMPLE NO. 1

Assume that an airplane has a gross weight of 20000 pounds. The runway altimeter indicates 3000 feet and the runway air temperature is 70°F. There is a 10 knot headwind. The runway length is 6500 feet and there is a 150 foot obstacle 8500 feet from the point where the ground run is begun. Enter the temperature scale at 70°F (21°C) and move up vertically to an altitude of 3000 feet and then horizontally over to a weight of 20000 pounds. Then move down to an obstacle height of 150 feet and horizontally left to the headwind correction grid. Follow the nearest guide line to a headwind velocity of 10 knots. Then move horizontally left again and read 8000 feet total distance to clear the obstacle. To obtain the ground run for this take-off after moving to the proper weight of 20000 pounds, continue vertically down without stopping at an obstacle height, to the vertical headwind grid. Follow the nearest guide line to a headwind of 10 knots and then move straight down and read a normal take-off run of 5900 feet.

EXAMPLE NO. 2

Assume the runway altimeter shows 5000 feet, the runway air temperature is 85°F, the headwind is 0, the runway length is 8500 feet, and there is a 150 foot obstacle 9500 feet from the start of the runway. Entering the curve at the temperature, and altitude it is seen that the recommended limit take-off weight is 18800 pounds. At this weight the ground run will be 7780 feet and the distance to clear 150 feet will be 11400 feet. The actual obstacle conditions of 150 feet at 9500 feet from start of runway, limit the take-off weight to 17900 pounds.

END

AN 01-85FGD-1
8 JUNE 1956

FLIGHT HANDBOOK INTERIM REVISION No. 17

Navy Model F9F-6-6P Aircraft

PUBLISHED BY DIRECTION OF THE CHIEF OF THE BUREAU OF AERONAUTICS

Of paramount interest to pilots. To be read by all pilots operating these aircraft

1. Cancellation: **None.**

2. Purpose: This interim instruction sets forth a recommended procedure for the use of oxygen in the subject aircraft.

3. Instructions: Add the following "NOTE" after the paragraph titled "Air Valve Control":

NOTE

To preclude adverse effects due to possible contamination, it is recommended that one hundred per cent (100%) oxygen be used on all flights of the subject aircraft as practicable considering the amount of oxygen available and duration of the flight.

AN 01-85FGD-1
27 APRIL 1956

FLIGHT HANDBOOK INTERIM REVISION No. 16

Navy Model F9F-6-6P Aircraft

PUBLISHED BY DIRECTION OF THE CHIEF OF THE BUREAU OF AERONAUTICS

Of paramount interest to pilots. To be read by all pilots operating these aircraft

1. Cancellation: None.

2. Purpose: This revision sets forth procedures to be followed in the event of inadvertent actuation of the firing mechanism of the Belt, Aircraft Safety, Pilot's, Automatic Release, Cartridge Actuated, MS16036.

3. Instructions: The MS16036 Belt Assembly was designed as a "one-shot" installation. In the event of inadvertent actuation of the firing mechanism with cartridge installed, the actuating assembly consisting of actuator assembly, MS16036-9, and actuating unit assembly, MS16036-10, will distort due to powder pressure and cannot be re-used. The following procedure shall apply:

   a. Remove MS16036 belt assembly from aircraft and place in RB category (Class 265) for repair.

   b. Install new MS16036 belt assembly procured from local stock.

END

INTERIM REVISION SUMMARY

The following Interim Revisions have been either canceled or incorporated in this Flight Handbook:

Canceled or Previously Incorporated

No. 2 (See CO 01-85FGD-1A)
No. 3 (See CO 01-85FGD-1A)
No. 4 Cancelled
No. 7 Cancelled
No. 10 (See CO 01-85FGD-1A)
No. 11 (See CO 01-85FGD-1A)
No. 12 (See CO 01-85FGD-1A)

Incorporated in This Revision on Pages Indicated

No. 1	105
No. 5	163, 164, 165
No. 6	36, 37
No. 8	58, 99, 100
No. 9	54, 78, 79, 101, 102, 103
No. 13	3, 85, 91

INTERIM REVISIONS OUTSTANDING: (To be maintained by custodian of Flight Handbook)

Number	*Date*	*Purpose*
NONE		